Aims and Scope

The series "Springer Theses" brings together a selection of the very best Ph.D. theses from around the world and across the physical sciences. Nominated and endorsed by two recognized specialists, each published volume has been selected for its scientific excellence and the high impact of its contents for the pertinent field of research. For greater accessibility to non-specialists, the published versions include an extended introduction, as well as a foreword by the student's supervisor explaining the special relevance of the work for the field. As a whole, the series will provide a valuable resource both for newcomers to the research fields described, and for other scientists seeking detailed background information on special questions. Finally, it provides an accredited documentation of the valuable contributions made by today's younger generation of scientists.

Theses may be nominated for publication in this series by heads of department at internationally leading universities or institutes and should fulfill all of the following criteria

- They must be written in good English.
- The topic should fall within the confines of Chemistry, Physics, Earth Sciences, Engineering and related interdisciplinary fields such as Materials, Nanoscience, Chemical Engineering, Complex Systems and Biophysics.
- The work reported in the thesis must represent a significant scientific advance.
- If the thesis includes previously published material, permission to reproduce this must be gained from the respective copyright holder (a maximum 30% of the thesis should be a verbatim reproduction from the author's previous publications).
- They must have been examined and passed during the 12 months prior to nomination.
- Each thesis should include a foreword by the supervisor outlining the significance of its content.
- The theses should have a clearly defined structure including an introduction accessible to new PhD students and scientists not expert in the relevant field.

Indexed by zbMATH.

More information about this series at http://www.springer.com/series/8790

Chung-Che Wu

Advanced and Applied Studies on Ultra-Trace Rare Earth Elements (REEs) in Carbonates Using SN-ICPMS and LA-ICPMS

Doctoral Thesis accepted by
National Taiwan University, Taipei, Taiwan

Springer

Author
Dr. Chung-Che Wu
Department of Chemistry
and Applied Biosciences
Swiss Federal Institute
of Technology (ETH)
Zürich, Switzerland

Supervisor
Dr. Chuan-Chou Shen
Department of Geosciences
National Taiwan University
Taipei, Taiwan

ISSN 2190-5053 ISSN 2190-5061 (electronic)
Springer Theses
ISBN 978-981-16-3618-9 ISBN 978-981-16-3619-6 (eBook)
https://doi.org/10.1007/978-981-16-3619-6

This Springer imprint is published by the registered company Springer Nature Singapore Pte Ltd.
The registered company address is: 152 Beach Road, #21-01/04 Gateway East, Singapore 189721, Singapore

Supervisor's Foreword

One of the major keys for understanding current and future climate changes is to reveal past climatic and environmental dynamics. Geochemical proxies, such as ultra-trace rare earth elements (REEs) in carbonates, have been used as a tracer to probe contemporary and quaternary climatic processes.

The main contribution of this dissertation is to develop frontier carbonate REE analytical techniques and apply them to decipher the impact of volcano eruptions on modern coral mortality and regional paleo-hydroclimate changes. Background, rationale, previous studies, and thesis structure were given in Chap. 1.

In Chap. 2, Dr. Wu presented his first developed world-leading analytical techniques on ultra-trace REE in carbonates using solution nebulization-inductively coupled plasma mass spectrometry (SN-ICPMS) in the High-Precision Mass Spectrometry and Environment Change Laboratory (HISPEC), Department of Geosciences, National Taiwan University. Our laboratory successfully demonstrated the capability of femtogram quantity carbonate REE analysis, carried out without chemical separation steps, and offered high-precision results (±1.9–6.5%, 2σ) (*Analytical Chemistry*, *83,* 6842–6848, 2011).

In Chap. 4, Dr. Wu developed a new direct carbonate REE analysis on laser ablation-inductively coupled plasma mass spectrometry (LA-ICPMS) in the Laboratory of Inorganic Chemistry, Department of Chemistry and Applied Biosciences, Swiss Federal Institute of Technology (ETH), under our collaborative partners, Dr. Detlef Günther and Dr. Bodo Hattendorf. This approach allows the direct determination of micro-domain stalagmite REEs containing low- to single-digit parts-per-billion (ppb) levels using LA-ICPMS (*Analytica Chimica Acta*, *1018,* 54–61, 2018).

Chapters 3 and 5 exhibit the application to natural carbonate corals on SN-ICPMS and stalagmites on LA-ICPMS, respectively. Combination of monthly resolved records of REEs/Ca and Al/Ca, and micro-domain images from corals in the South China Sea suggests that the coral mortality event in 1991 was exacerbated by heavy ash fallout from the cataclysmic 1991 volcanic eruption of Mount Pinatubo. This study highlighted the profound impact of a volcanic eruption on the modern vulnerable coral reef ecosystem under the stress of global warming (*Geophysical Research*

Letters, 45, 12396–12402, 2018). Dr. Wu also displayed the feasibility for the identification of anomalous climate/volcano events at 88–78 thousand years ago by substantial REE anomalies of 100 s ng/g and associated Al pulses of 100 s μg/g in a fossil stalagmite from East Timor. This study gave the potential and the applicability to understand the impacts of volcano eruptions on biosystem and paleoclimate regimes.

In summary, developments of high-quality ultra-trace carbonate REE analytical techniques and related applications described in this dissertation can contribute to diverse fields in oceanic and geological societies. I believe that the techniques and findings can potentially inspire students and junior researchers to work on development of frontier analytical methods and unexplored topics.

Taipei, Taiwan
March 2021

Dr. Chuan-Chou Shen

Abstract

Two state-of-the-art quantitative techniques to determine ultra-trace rare earth elements (REEs) in natural carbonates using solution nebulization-inductively coupled plasma mass spectrometry (SN-ICPMS) and laser ablation-inductively coupled plasma mass spectrometry (LA-ICPMS) with respective applications were presented in this dissertation. These techniques were applied to natural carbonates, including corals and stalagmites, to understand volcano eruptions and the impacts on modern biosystem and paleoclimate regimes.

In the first SN-ICPMS protocol, direct measurements for femtogram quantity carbonate samples without chemical separation steps can offer accurate and high-precision analysis ($\pm$1.9–6.5%, 2σ) with a high sample throughput of 8–10 samples/hr routinely (*Analytical Chemistry*, 2011). Application to modern *Porites* corals collected from South China Sea region, the anomalies of REE contents and Al/Ca ratios associated with micro-domain images, register modern coral reefs could be exacerbated by volcanic eruptions (*Geophysical Research Letters*, 2018).

In the second protocol, a high-sensitivity quantitative open-cell LA-ICPMS technique has been established to allow direct sampling on stalagmite surface in the atmospheric air. This technique improved limits of detection down to sub-ng/g range and promises analyses of carbonate REE profiles at the single-digit parts-per-billion (ppb) levels (*Analytica Chimica Acta*, 2018). Application to a 15-cm stalagmite collected from East Timor reveals two peaks of REE contents by at least one order of magnitude, possibly due to volcanic ash preserved in stalagmite.

Both improved SN-ICPMS and LA-ICPMS techniques highlight the high-sensitivity and high-temporal-resolution carbonate REE analyses for corals and stalagmites, with great potential to other natural carbonates such as travertine, tufa, and flowstone, and benefit our understanding of paleoclimatic and paleoenvironmental dynamics.

Keywords Carbonates · Rare earth elements · ICPMS

Parts of this thesis have been published in the following journal articles:

1. Shen CC, Wu CC, Liu Y, Yu J, Chang CC, Lam DD et al (2011) Measurements of natural carbonate rare earth elements in femtogram quantities by inductive coupled plasma sector field mass spectrometry. Anal Chem 83(17):6842–68482.
2. Wu C-C#, Burger M#, Günther D, Shen C-C, Hattendorf B (2018) Highly-sensitive open-cell LA-ICPMS approaches for the quantification of rare earth elements in natural carbonates at parts-per-billion levels. Anal Chim Acta 1018:54–61. #These authors contributed equally to this work.
3. Wu C-C, Shen C-C, Lo L, Hsin Y-C, Yu K, Chang C-C et al (2018) Pinatubo volcanic eruption exacerbated an abrupt coral mortality event in 1991 summer. Geophys Res Lett 45:12396–12402.

Acknowledgements

This work would not have been achieved without the help of a number of people, and I would like to acknowledge them here:

First, I would like to express my sincere gratitude to my advisor Dr. Chuan-Chou Shen, for the advice and support of my PhD study and research. I would also thank my all committee members, Dr. Chung-Ho Wang, Dr. Sheng-Rong Song, Dr. Hong-Chun Li, Dr. Hao-Yang Lee, and Dr. Kuo-Fang Huang, for those constructive comments and suggestions they gave me to improve this dissertation.

Besides, I would like to thank Prof. Dr. Detlef Günther who provided me the precious opportunity to carry out one of my PhD projects in his laboratory. I thank Dr. Bodo Hattendorf for his guidance, advice, and the confidence he gave to me. I thank Dr. Marcel Burger and all the members in the Günther research group, for their kind help while I was in ETH, Switzerland. I missed those old good days in Zurich.

Deep gratitude is due to my parents who have always supported and prayed for me unconditionally. In particular, I would like to thank my wife, Chia-Ying, who has always showed understanding and was supportive when I doubted myself. Without the courage she gave me, my studies and doctorate would not have been possible.

I also thank the financial supports: one was from Taiwan ROC MOST which allowed me to perform an overseas project (104-2917-I-002-007); another was the European Union's Horizon 2020 under the Marie Skłodowska-Curie grant agreement (No. 891710).

As being a Christian, I believe this a road that the Lord leads me, though it is the one less travelled by. All I can do is to lean on Him and trust Him. I believe, then, it will make all the difference.

but, a vast amount of material, 5–10 g per sample, was required [22]. With the advance of solution nebulization-inductively coupled plasma spectrometry (SN-ICPMS) techniques, multi-element analysis for carbonate REEs in mg/g to ng/g levels were possible [1, 2, 27]. In 2011, I developed a fast and precise SN-ICPMS technique for determination of carbonate REEs in femtogram quantities without chemical separation process [24]. Although a high sample throughput of 8–10 samples/hr could be achieved, steps of physical subsampling, dissolution, and dilution were still required. Using SN-ICPMS approach remains limited by a discrete mini-domain sampling strategy and time-consuming chemical preparation procedures.

1.3 Advantages of Using LA-ICPMS Approach to Determine Carbonate REE Profiles

Laser ablation-inductively coupled plasma mass spectrometry (LA-ICPMS) techniques can beneficially serve the strengths of rapid, *in-situ*, and high lateral scanning to micro-domain sample specimen [10]. Since the 2000s, LA-ICPMS has become one of the widely applied techniques for solid sample analyses in a variety of fields, such as geology [6, 30], biology [14, 32], archaeology [15, 18], and material sciences [11]. It offers rapid and direct analyses at a lateral resolution of 1–100 s µm for the determination of major, minor, trace, ultra-trace elements and isotopic compositions [8, 13]. Other advantages include quantitative multi-element capabilities and a large dynamic range [10]. However, it was still a challenge to quantify carbonate REEs at parts-per-billion (ppb, ng/g) levels using LA-ICPMS.

1.4 Structure of This Dissertation and Short Summaries of Results

In the Chap. 2, a novel approach using SN-ICPMS allows directly measuring femtogram quantity carbonate samples without chemical separation steps has been established. In the Chap. 3, this approach was applied to materials of modern *Porites* corals collected from South China Sea (SCS). The results indicate that an intense coral mortality event occurred in 1991 in SCS region could be linked to the giant volcanic eruption of Mount Pinatubo. Following the Chap. 4, the second novel approach was presented. An advanced LA-ICPMS method allows to quantitative determine stalagmite REE concentrations in ng/g range. In the Chap. 5, high resolution REE and Al profiles of a stalagmite collected from East Timor reveal at least two abrupt elevations in the time period of 78–88 thousand years ago (ka BP, before 1950 AD). Volcanic eruptions might be the main contributors to these anomalies.

In the Chap. 6, summarized and perspective points are raised to this dissertation. Developing advanced analytical approaches can contribute to our further understanding of paleoclimatic and paleoenvironmental changes, as well as profit the projection of future climate changes.

References

1. Akagi T, Hashimoto Y, Fu FF, Tsuno H, Tao H, Nakano Y (2004) Variation of the distribution coefficients of rare earth elements in modern coral-lattices: species and site dependencies. Geochim Cosmochim Acta 68(10):2265–2273
2. Baker J, Waight T, Ulfbeck D (2002) Rapid and highly reproducible analysis of rare earth elements by multiple collector inductively coupled plasma mass spectrometry. Geochim Cosmochim Acta 66(20):3635–3646
3. Chu G, Sun Q, Wang X, Liu M, Lin Y, Xie M et al (2012) Seasonal temperature variability during the past 1600 years recorded in historical documents and varved lake sediment profiles from northeastern China. The Holocene 22:785–792
4. Dai R, Liu C, Xiao B (2007) Chinese character recognition: history, status and prospects. Front Comput Sci China 1(2):126–136
5. DeLong KL, Quinn TM, Taylor FW, Lin K, Shen CC (2012) Sea surface temperature variability in the southwest tropical Pacific since AD 1649. Nat Clim Chang 2(11):799
6. Fallon SJ, White JC, McCulloch MT (2002) *Porites* corals as recorders of mining and environmental impacts: Misima Island, Papua New Guinea. Geochim Cosmochim Acta 66(1):45–62
7. Fairchild IJ, Smith CL, Baker A, Fuller L, Spötl C, Mattey D, McDermott F (2006) Modification and preservation of environmental signals in speleothems. Earth Sci Rev 75(1–4):105–153
8. Günther D, Hattendorf B (2005) Solid sample analysis using laser ablation inductively coupled plasma mass spectrometry. TrAC Trends Anal Chem 24(3):255–265
9. Haley BA, Klinkhammer GP, Mix AC (2005) Revisiting the rare earth elements in foraminiferal tests. Earth Planet Sci Lett 239(1–2):79–97
10. Hattendorf B, Günther D (2014) Laser ablation inductively coupled plasma mass spectrometry (LA-ICPMS). In: Handbook of spectroscopy, Second, Enlarged Edition, pp 647–698
11. Herman PR, Marjoribanks RS, Oettl A, Chen K, Konovalov I, Ness S (2000) Laser shaping of photonic materials: deep-ultraviolet and ultrafast lasers. Appl Surf Sci 154:577–586
12. Jaiswal KS, Wald DJ, Earle PS, Porter KA, Hearne M (2011) Earthquake casualty models within the USGS prompt assessment of global earthquakes for response (PAGER) system. In: Human casualties in earthquakes. Springer, Dordrecht, pp 83–94
13. Jochum KP, Scholz D, Stoll B, Weis U, Wilson SA, Yang Q et al (2012) Accurate trace element analysis of speleothems and biogenic calcium carbonates by LA-ICP-MS. Chem Geol 318:31–44
14. Konz I, Fernández B, Fernandez ML, Pereiro R, Sanz-Medel A (2012) Laser ablation ICP-MS for quantitative biomedical applications. Anal Bioanal Chem 403(8):2113–2125
15. Lazic V, Colao F, Fantoni R, Spizzicchino V (2005) Laser-induced breakdown spectroscopy in water: improvement of the detection threshold by signal processing. Spectrochim Acta Part B 60(7–8):1002–1013
16. Liu KB, Shen C, Louie KS (2001) A 1,000-year history of typhoon landfalls in Guangdong, southern China, reconstructed from Chinese historical documentary records. Ann Assoc Am Geogr 91(3):453–464
17. Liu Y, Lo L, Shi Z, Wei KY, Chou CJ, Chen YC et al (2015) Obliquity pacing of the western Pacific Intertropical Convergence Zone over the past 282,000 years. Nat Commun 6:10018
18. Neff H (2003) Analysis of mesoamerican plumbate pottery surfaces by laser ablation-inductively coupled plasma-mass spectrometry (LA-ICP-MS). J Archaeol Sci 30(1):21–35

19. Palmer MR (1985) Rare earth elements in foraminifera tests. Earth Planet Sci Lett 73(2–4):285–298
20. Reynolds RW, Rayner NA, Smith TM, Stokes DC, Wang W (2002) An improved in situ and satellite SST analysis for climate. J Clim 15(13):1609–1625
21. Richter DK, Gotte T, Niggemann S, Wurth G (2004) REE^{3+} and Mn^{2+} activated cathodoluminescence in late glacial and Holocene stalagmites of central Europe: evidence for climatic processes? The Holocene 14(5):759–768
22. Scherer M, Seitz H (1980) Rare-earth element distribution in Holocene and Pleistocene corals and their redistribution during diagenesis. Chem Geol 28:279–289
23. Shen CC, Lin K, Duan W, Jiang X, Partin JW, Edwards RL et al (2013) Testing the annual nature of speleothem banding. Sci Rep 3:2633
24. Shen CC, Wu CC, Liu Y, Yu J, Chang CC, Lam DD et al (2011) Measurements of natural carbonate rare earth elements in femtogram quantities by inductive coupled plasma sector field mass spectrometry. Anal Chem 83(17):6842–6848
25. Siklósy Z, Demeny A, Vennemann TW, Pilet S, Kramers J, Leél-Őssy S et al (2009) Bronze Age volcanic event recorded in stalagmites by combined isotope and trace element studies. Rapid Commun Mass Spectrom 23(6):801–808
26. Tan L, Shen CC, Cai Y, Lo L, Cheng H, An Z (2014) Trace-element variations in an annually layered stalagmite as recorders of climatic changes and anthropogenic pollution in Central China. Quatern Res 81(2):181–188
27. Webb GE, Kamber BS (2000) Rare earth elements in Holocene reefal microbialites: a new shallow seawater proxy. Geochim Cosmochim Acta 64(9):1557–1565
28. Wu C-C, Shen C-C, Lo L, Hsin Y-C, Yu K, Chang C-C et al (2018) Pinatubo volcanic eruption exacerbated an abrupt coral mortality event in 1991 summer. Geophys Res Lett 45:12396–12402
29. Wu C-C, Burger M, Günther D, Shen C-C, Hattendorf B (2018) Highly-sensitive open-cell LA-ICPMS approaches for the quantification of rare earth elements in natural carbonates at parts-per-billion levels. Anal Chim Acta 1018:54–61
30. Wyndham T, McCulloch M, Fallon S, Alibert C (2004) High-resolution coral records of rare earth elements in coastal seawater: biogeochemical cycling and a new environmental proxy. Geochim Cosmochim Acta 68(9):2067–2080
31. Zhou H, Wang Q, Zhao J, Zheng L, Guan H, Feng Y, Greig A (2008) Rare earth elements and yttrium in a stalagmite from Central China and potential paleoclimatic implications. Palaeogeogr Palaeoclimatol Palaeoecol 270(1–2):128–138
32. Zoriy MV, Kayser M, Izmer A, Pickhardt CBJS, Becker JS (2005) Determination of uranium isotopic ratios in biological samples using laser ablation inductively coupled plasma double focusing sector field mass spectrometry with cooled ablation chamber. Int J Mass Spectrom 242(2–3):297–302

Chapter 2
Approach I: Rapid and Precise Measurements of Natural Carbonate Rare Earth Elements in Femtogram Quantities by Solution Nebulization-Inductively Coupled Plasma Mass Spectrometry (SN-ICPMS)

2.1 Introduction

Since the 1980s, rare earth elements (REEs) have emerged as important tracers in diverse applications of the earth sciences, including evolution of lithospheric reservoirs, modern and past environmental changes, and marine geochemistry and ocean circulation (e.g., [5, 23, 32, 3, 37]). Records of REEs in natural carbonate materials, such as corals [6, 22, 30, 35], foraminifera [9, 17], and speleothems [31, 39], have been analyzed for understanding contemporary and late-Quaternary climatic and environmental changes. However, overall applicability has been sharply limited due to the low REE abundances, 1–100 s nmol/mol, in most natural samples and stringent analytical difficulties.

Different analytical techniques, such as neutron activation analysis (NAA; [22]), inductively coupled plasma-atomic emission spectrometry (ICP-AES; [12]), inductive coupled plasma mass spectrometry (ICPMS; [1, 33]), isotope dilution-thermal ionization mass spectrometry (ID-TIMS; [17, 30]), cathodoluminescene [19], and laser ablation-ICPMS (LA-ICPMS; [6, 35]), have been employed for carbonate REE determinations. ID-TIMS can give good precision of 0.1–3% (2 relative standard deviation, 2 RSD) [30], however, labor-intensive sample preparation processes limit the rate of analytical measurements. With improvements in instrumentation, including high sensitivity, low detection limit, and rapid multi-elements analysis in the past decades, ICPMS provides the high ability to analyze carbonate REEs with precision of 6–20% (e.g., [33, 1]). However, for both ID-TIMS and ICPMS, time-consuming column chromatography is required to separate REEs from a matrix and

Electronic supplementary material The online version of this chapter (https://doi.org/10.1007/978-981-16-3619-6_2) contains supplementary material, which is available to authorized users.

C.-C. Wu, *Advanced and Applied Studies on Ultra-Trace Rare Earth Elements (REEs) in Carbonates Using SN-ICPMS and LA-ICPMS*, Springer Theses,
https://doi.org/10.1007/978-981-16-3619-6_2

this step ultimately hinders sample throughput. Simple ICPMS methods (e.g., [4, 39]) using enriched isotopes and elemental internal standards and without chemical separation steps give a 2 RSD of ±10–100% for nmol/mol-level REE analyses. Modern LA-ICPMS techniques can offer a direct and fast measurement of micrometer-resolution carbonate REE concentrations (e.g., [6, 35]). Due to low REE abundances in corals, for example, 2 RSD of ±26–36% was achieved [35].

In this study, protocols to directly measure carbonate REE abundances by solution nebulization-ICPMS (SN-ICPMS) with a 2 RSD reproducibility of 1.9–6.5% for 10–20 μg carbonate samples have been established. Following factors affecting the high-precision determination of REE/Ca ratios were carefully addressed and solved: (1) spectral interferences [10, 18](2) mass discrimination and ratio drifting (e.g., [21, 27]), and (3) chemical matrix effects [21, 26]. Examples include a living *Porites* coral core, planktonic foraminifer *Globorotalia menardii*, and benthic foraminifer *Cibicidoides wuellerstorfi*.

2.2 Experimental Section

2.2.1 Reagents, Standards, and Samples

Preparation of standards and samples were performed in a class-10,000 geochemical clean room with class-100 benches in the High-precision Mass Spectrometry and Environment Change Laboratory (HISPEC), Department of Geosciences, National Taiwan University. Water was purified using an ultrapure water tandem system with Millipore Milli-Q ACADEMIC and Milli-Q ELEMENT. Teflon and polyethylene vials, bottles and beakers were cleaned by boiling with 3 N guaranteed regent grade (GR) HNO_3 (Merk & CO. Inc.) for least 4 h. Ultrapure reagents from SEASTAR or J.T. BAKER were used in chemistry.

One in-house matrix-matched standard, CarbREE-I (REE/Ca ratios of 464.5–368.8 nmol/mol, Table 2.S1) was gravimetrically prepared with super-pure calcium carbonate powder (purity ≥99.999%, Sigma-Aldrich Inc.) and REE solution standard (La, Ce, Pr, Nd, Sm, Eu, Gd, Tb, Dy, Ho, Er, Tm, Yb, Lu; 10 μg/ml, High-Purity Standards) in 5% ultrapure HNO_3.

Natural carbonate standards were prepared. Two thousand individuals of a calcitic planktonic foraminifer, *G. menardii* from a marine sediment core ODP1115B (09° 110′ S, 151° 340′ E, water depth 1148.8 m), were cleaned and dissolved to provide a foraminiferal reference solution, FORAM-GM [27]. A 0.5 cm-thick sectioned slab of a living massive *Porites* coral head ST0506, collected offshore central Vietnam (16° 13′ N, 108° 12′ E) in 2005 [28], was cut along the growth direction and cleaned [24]. One 0.1 g bulk sample was cut from the 1991 band, cleaned [25], and dissolved in 5% HNO_3. The foraminifer FORAM-GM and coral ST0506 solutions as well as CarbREE-I were used for assessing analytical reproducibility. Due to the lack of an adequate certified reference material for REE in carbonate samples, an REE-admixed coral standard was applied to validation of the proposed technique. This coral REE standard, CoralM-REE, with theoretical REE/Ca ratios of 33–56 nmol/mol was made by gravimetrically mixing the REE solution standard and one in-house coral standard solution with low REE/Ca ratios of 0.15–5.9 nmol/mol, CoralM, prepared with a modern *Porites* coral collected in Nanwan (21°57′ N, 120°45′ E), Taiwan in 2003. It was used for accuracy evaluation.

Coral and foraminiferal samples were also used in this study. Interlaboratory comparison between quadrupole-ICPMS (Q-ICPMS) [20, 36] and our SF-ICPMS methods was performed by analyzing REE/Ca ratios of a calcitic benthic foraminifer *C. wuellerstorfi* selected from a depth interval of 156–157 cm in a gravity core MW91-9 GGC-15 (0° N, 158° E, water depth 2310 m), located on Ontong Java Plateau. A modern *Porites* coral core, WZI-1, 20 cm in length and 5 cm in diameter, with a growth rate of 7–8 mm/year, was drilled offshore of Weizhou Island (WZI) (21° 10' N, 109° 40′ E) in the northern SCS in 2009 (Fig. 2.S1 of the **Supplementary information**). Subsamples, 2–4 mg each, were cut on a sliced slab at an interval of 0.7 mm from 2002 to 2005 for monthly resolved REE/Ca determination.

2.2.2 *Safety Considerations*

Nitric acid is a toxic, corrosive reagent that can burn skin and damage respiratory organs. A fume hood with goggles and protective gloves are required to avoid inhalation and contact with skin and eyes. Eye-wash stations and safety showers should be available in case of accidental exposure. Acidic solutions should be neutralized prior to disposal.

2.2.3 *Instrumentation*

Measurements were carried out on a Finnigan Element II SF-ICPMS (Thermo Electron, Bremen, Germany) at low resolution ($M/\Delta M = 300$). Radio frequency power was set at 1200 W. Argon flow rates were set at 16 L/min for the plasma gas, 0.8–1.2 μL/min for the auxiliary gas, and 0.8–1.0 L/min for the sample gas. The Aridus dry introduction system (CETAC Technologies, NE) with a sample solution uptake rate of 80 μL/min was used. The daily optimum condition was 4–6 L/min for sweep Ar flow and 0.05–0.15 L/min for N_2 flow. The temperatures of the spray chamber and desolvator were set at 110 °C and 160 °C, respectively. This system provided a 5 to tenfold enhancement in sensitivity and dramatically reduced the polyatomic inferences from hydrides and oxides [26]. The ASX-100 Micro Autosampler (CETAC Technologies, NE) was applied for automatic sequence measurements.

A single secondary electron multiplier in peak-hopping mode was used to measure the ion beams of ^{46}Ca, ^{138}Ba, ^{139}La, ^{140}Ce, ^{141}Pr, ^{146}Nd, ^{147}Sm, ^{153}Eu, ^{159}Tb, ^{160}Gd,

^{163}Dy, ^{165}Ho, ^{166}Er, ^{169}Tm, ^{172}Yb, and ^{175}Lu. Ion beam intensities of $^{46}Ca^+$ of 0.8–1.5 × 10^6 cps before running samples, and instrumental drift was calibrated by measuring bracketed standard solutions between samples. Magnetic-scan (B-scan) mode was used for peak jumping between masses 46, 138, and 159, and electrostatic-scan (E-scan) mode for masses of 139–153 and 159–175. For each measurement, sample solution uptake lasted for 190 s, followed by a 70-s washout step with 5% HNO_3. Every four samples were bracketed with one standard. All REE/Ca ratios were calculated directly from ratios of ion beam intensities using external matrix-matched standards to correct for blanks, instrumental mass discrimination, and ratio drifting. Data were calculated in an off-line data reduction process, modified from Shen et al. [26]. All errors given are two standard deviations (2σ) or 2 RSD unless otherwise noted. Detailed instrumental settings and data acquisition methods are summarized in Tables 2.S2 and 2.S3.

2.3 Results and Discussion

2.3.1 Blanks and Spectral Interferences

The procedural blank (PB), including chemical blank and spectral interferences, is <2 × 10^3 cps at 46 atomic mass unit (amu) and <70 cps at 139–175 amu. The PB/signal ratio is 0.1–0.2% for ^{46}Ca and <0.14% for all REE isotopes of interest in CarbREE-I standard solution with 20–40 ppm Ca. For natural carbonates with <10 s nmol/mol REEs, the background/signal ratio is 2–3% for ^{153}Eu, ^{159} Tb, ^{160}Gd, and ^{163}Dy, and <1% for other measured REE isotopes. There are two types of instrumental blank in the SN-ICPMS. One is residual signal from previous samples. After uptaking 5% HNO_3 for 70s, the calcium and REE blanks could be reduced to <0.02–0.03% of original signals. These low blanks were subtracted from measured ion beams to earn net intensities. Another type of memory interference is "ghost" ion beam due to a significant contamination after running high concentration calcium for 1–2 h. These unexpected abrupt spiky signals were filtered in an off-line data reduction process [26].

Serious isobaric interference of oxide and hydroxide forms of Ba on Eu was reported for wet introduction systems (e.g., [10, 18]). The barium oxide/barium ratio is only 0.004% in our method with a desolvation introduction system (Fig. 2.S2). Ba/Ca ratios range about 1–10 μmol/mol (e.g., [13, 2]) in marine carbonates of corals and foraminifera,and the BaO^+ is <50 cps at 154 amu and <10 cps at 153 amu. The correspondent isobaric interference for $^{153}Eu^+$ is <2.5% for samples with low REE levels of 1–10 nmol/mol. The $^{137}BaO^+$ can cause a high isobaric interference, ~ 20% of $^{153}Eu^+$ ion beam, for carbonates with Ba/Ca level >100 μmol/mol and Eu/Ca < 10 nmol/mol. Hence, $^{138}Ba^+$ ion beam was measured and the isobaric interference was corrected using an intensity ratio of 10.7 × 10^6 $[(^{137}BaO^+ + {}^{136}BaOH^+)/{}^{138}Ba^+]$

for all carbonates. $^{140}Ce^{16}O^+$ is an isobaric interference for $^{156}Gd^+$ measurement [7]. To avoid this spectral interference, $^{160}Gd^+$ was selected to determine Gd content.

2.3.2 Mass Discrimination and Isotopic Ratio Drifting

Increased ion transmission with mass causes deviations of measured ratios from true ratios and this mass discrimination (or mass bias) affects analytical precision and accuracy in ICPMS (e.g., [21, 26, 27]). Low-frequency drift of measured isotopic ratios is also observed (Fig. 2.1a) mainly due to the deposition of high-calcium samples, various plasma conditions, and drift in the electronics (e.g., [21, 27]). One typical example of low-frequency drift of measured La/Ca ratios is shown in Fig. 2.1b. La/Ca ratios can vary 20% in 2 h. To correct for these types of biases, the CarbREE-I standard solution was measured between every 4 samples. After linear-interpolation correction with a bracketing standard, a short-term external 2 RSD precision of 0.8% for La/Ca can be achieved (Fig. 2.1b).

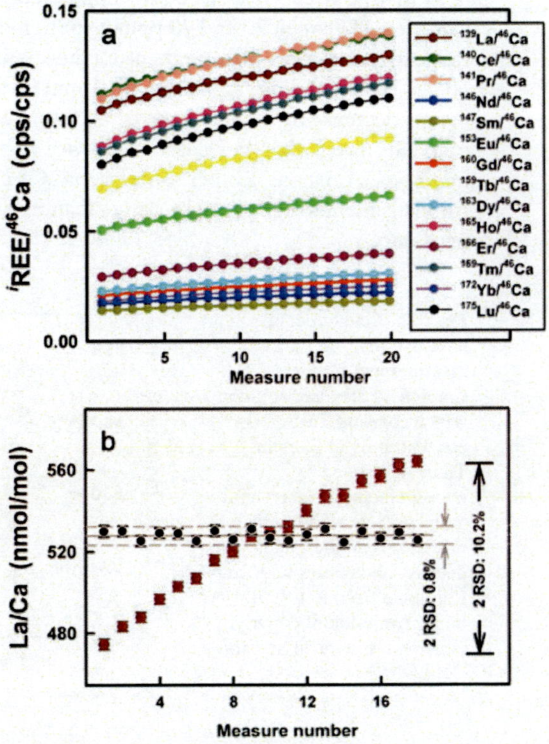

Fig. 2.1 **a** Mass drifting in measured REE isotope (^{i}REE)/^{46}Ca ratios over a two-hour analysis for the 20 ppm-Ca CarbREE-I standard solution. **b** Ratio drift in La/Ca over a two-hour measurement experiment. Short-term 2 RSD external precision is improved from ±10% (dark red circles) for the uncorrected values to ±0.8% (black circles) for the standard-bracketing corrected data. Reprinted from Ref. [29] with permission of copyright American Chemical Society

2.3.3 *Matrix Effects*

The acid effect, one type of one kind of matrix effect observed in ICPMS, was evaluated using wet introduction systems with ion beam intensities of different elements ([34, 15]). As an example the La/Ca ratios of a 20 ppm-Ca CarbREE-I standard solution in different strength HNO_3 ranging 1–6% are plotted in Fig. 2.2. The measured La/Ca ratios for 6% HNO_3 are slightly higher than the other values, showing a minor acid effect. The external errors of 1.6–2.4% for 1–3% HNO_3 are 1.5–2 times larger than the internal errors of 1.1–1.3%. The small external errors for high HNO_3 levels of 4–6% are comparable to the internal precisions. A compensative 5% HNO_3 with good-reproducible data was determined to be acceptable for routine measurements.

Previous studies (e.g., [8, 27]) have shown a matrix effect of Ca concentration on accurate elemental ratio determinations. This effect was addressed by measuring REE/Ca ratios in different diluted standard solutions with Ca levels of 5–50 ppm. Figure 2.3 shows the Ca matrix effect on La/Ca and Yb/Ca. For [Ca] of 5–20 ppm, measured La/Ca ratio decreases with a slope of ~17 La/Ca (nmol/mol)/[Ca] (ppm). For [Ca] $\geq$ 20 ppm, the Ca matrix effect is trivial (Fig. 2.3). A similar Ca matrix effect on Yb/Ca measurement with decreasing trends is observed. To attenuate the [Ca] matrix effect, [Ca] in both sample and standard aliquots was adjusted to the same level between 20 and 40 ppm prior to measurement. One disadvantage for this fast carbonate REE measurement method is inevitable Ca deposition on sampling cones. When running samples with 40 ppm Ca concentration, the intensities of the ion beams decrease by a typical rate of ~10%/h. The sampling cones should be cleaned manually every day. For carbonates with relatively high REE/Ca ratios of tens to hundreds of nanomoles per mole, a low Ca concentration of 20 ppm is suggested for instrumental analysis. The daily cleaning step for the sampling cones will not be necessary.

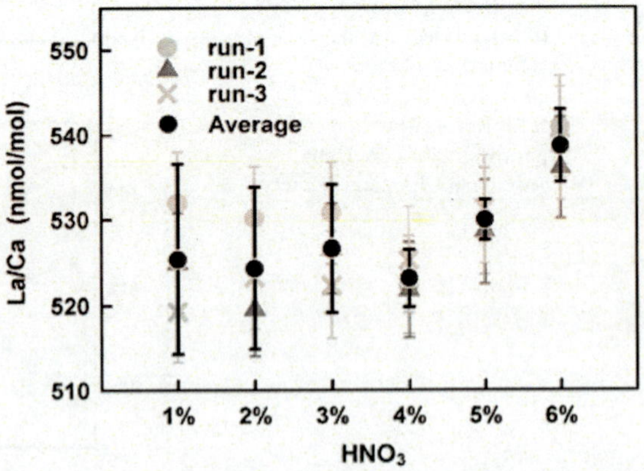

Fig. 2.2 Acid effect of HNO_3, 1–6%, on La/Ca measurement for the CarbREE-I standard solution with a constant Ca concentration of 20 ppm. Three duplicated measurements are shown in gray symbols with internal precision, and the average data as solid circles with 2σ. Reprinted from Ref. [29] with permission of copyright American Chemical Society

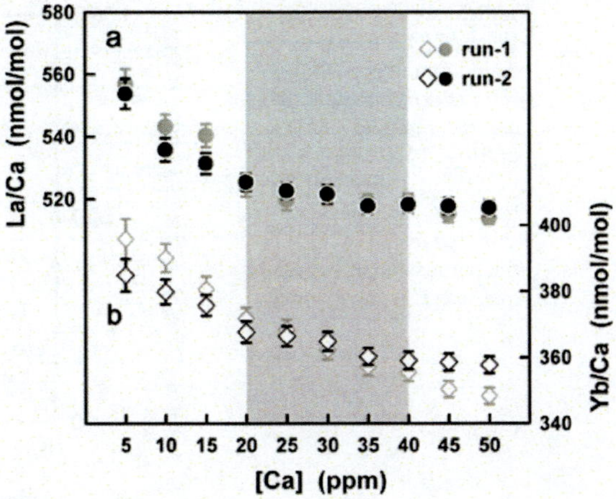

Fig. 2.3 Duplicated runs for Ca concentration effect on determinations of **a** La/Ca and **b** Yb/Ca. There is no significant effect of Ca levels over 20–40 ppm (shaded area). Reprinted from Ref. [29] with permission of copyright American Chemical Society 2011

2.3.4 Accuracy

REE concentration uncertainties in the commercially available REE solution standard are 1.0% (2 RSD), based on the reports given by High-Purity Standards (Charleston, SC). The use of super pure Sigma-Aldrich calcium carbonate powder contributes 13.6% (±1.4%) for La, 35.5% (±3.0%) for Ce, and <0.3% for all other REEs in the gravimetrically prepared CarbREE-I solution standard. The uncertainties of true REE concentrations in this CarbREE-I standard, including impurities from the calcium carbonate powder, are estimated to be ±1.5% for La, ±3.2% for Ce, and ±1.0% for Pr, Nd, Sm, Eu, Gd, Tb, Dy, Ho, Er, Tm, Yb, and Lu. Measured low CarbREE-II REE/Ca ratios determined with CarbREE-I are insignificantly different from gravimetric values (Table 2.S1), suggesting a good linearity of intensity signals.

The differences between determined REE/Ca ratios and gravimetrically calculated values in the admixed coral standard solution, CoralM-REE, are all less than the long-term 2 RSD uncertainties of 2–6% for coral carbonate (Fig. 2.4). This indicates there

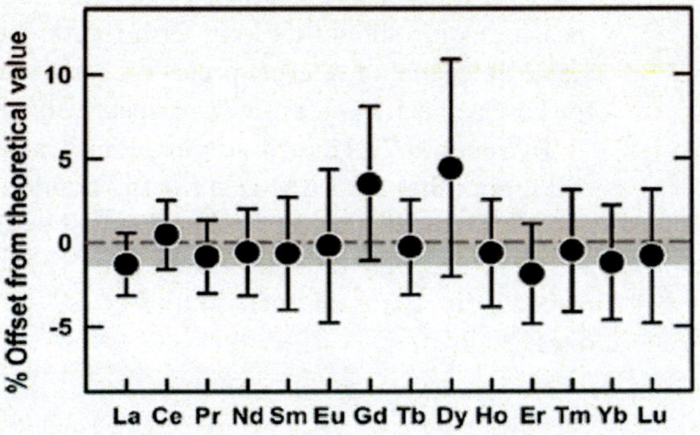

Fig. 2.4 Percent offset of measured REE/Ca ratios with long-term 2 RSD error bars (Table 2.S4) from theoretical values in CoralM-REE standard solution. Shaded area denotes the uncertainty of REE/Ca in the standard. Reprinted from Ref. [29] with permission of copyright American Chemical Society 2011

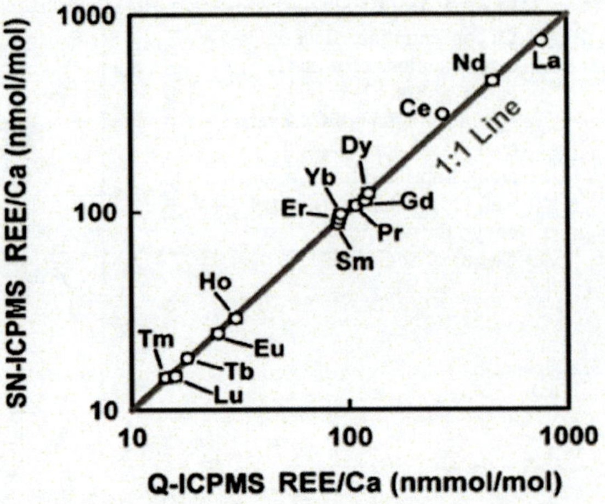

Fig. 2.5 Comparison of measured REE/Ca ratios of a benthic foraminifer, *C. wuellerstorfi*, from marine core MW91-9 GGC-15 by our SN-ICPMS and Q-ICPMS [20, 36] at the University of Cambridge. Reprinted from Ref. [29] with permission of copyright American Chemical Society

is no significant matrix effect to quantify REE contents in natural carbonates by using the in-house standards. An interlaboratory comparison for a benthic foraminifer *C. wuellerstorfi* sample from core GGC-15 shows that our SN-ICPMS results replicate measurements made by Q-ICPMS [20, 36] at the University of Cambridge (Fig. 2.5), with the exception of Ce/Ca. This sole discrepancy (13 $\pm$ 4% for Ce/Ca) is attributed to slightly heterogeneous Ce contents between the two subsamples that were used in the two laboratories. The abovementioned experiments indicate our SF-ICPMS methods can accurately be applied to natural carbonates without a chemistry separation process.

2.3.5 *Precision and Detection Limit*

Analytical reproducibility over the course of 6 months (Dec 2008–May 2009) determined by repeatedly analyzing the CarbREE-I standard solution is plotted in Fig. 2.6a. Internal 2 RSD precision of REE/Ca ratios ranges from 0.4–1.4%. Replicate measurements show that long-term reproducibility ranges 0.8–1.7% (Table 2.S4), 1.5–2 times that of internal precision. An external uncertainty, for example, is 0.9% for La/Ca and 1.1% for Yb/Ca, respectively (Fig. 2.6a).

Two-month (2010 July–August) REE/Ca data of coral ST0506 (2–242 nmol/mol) and foraminifer FORAM-GM (3–157 nmol/mol) are given in Table 2.S4 and La/Ca and Yb/Ca are also illustrated in Fig. 2.6b, c. Internal 2 RSD precision ranges from 1 to 2% for high (10–100 s nnmol/mol) and low (<10 nnmol/mol) REE levels, respectively. External uncertainties of 1.9–6.5% are 2–4 times larger than internal ones.

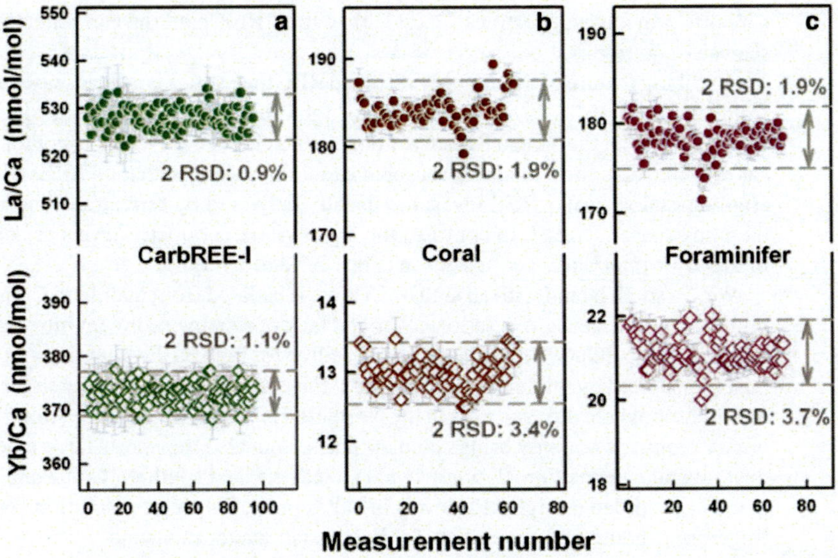

Fig. 2.6 Replicate measurements of La/Ca (solid symbols) and Yb/Ca (hollow symbols) with internal precision of measured REE/Ca ratios in **a** CarbREE-I, **b** coral ST0506, and **c** foraminifer FORAM-GM. External uncertainties are given with dashed line and values in gray. Reprinted from Ref. [29] with permission of copyright American Chemical Society

Large external errors of 5–6% are observed for Eu, Gd, and Dy in coral ST0506 (Table 2.S4). Relatively high 2σ external deviations of 5–6% and positive offsets of 3–4% are also observed for Gd and Dy in the reference sample CoralM-REE (Fig. 2.4). The high deviations and offsets are likely caused by polyatomic interferences from organics and/or complexes, indicating matrix complexity of natural carbonates.

On the basis of the definition of the detection limit, $3 \times \delta_B$, where δ_B is the standard deviation of the procedural background on ICPMS (10–40 cps), the calculated detection limits of carbonate REE/Ca ratios range 0.1–0.4 nnmol/mol. The Eu detection limit varies with Ba content on ICPMS (see section "Blanks and spectral interferences"). The isobaric spectral background should be well-estimated, or the Eu/Ca ratio will be biased.

2.3.6 Monthly Coral REE Records

The WZI-1 coral REE diagram shows 4-year monthly averages (AD 2002–2005), normalized to the post-Archean Australian shale (PAAS) [14] and is characterized by heavy REE (HREE) enrichment and a negative Ce anomaly (Fig. 2.S2). The two main features resemble those of SCS surface seawater (Fig. 2.S3) [16]. This observation supports previous suggestions [30, 33] that coral REE patterns can reflect ambient seawater conditions.

A striking feature of the monthly resolved REE time series is a clear seasonal cycle, from low levels of 40–80 ppb in summer to high values of 120,140 ppb in winter/early spring (Fig. 2.7). Individual REE/Ca ratios, even as low as 0.3–1 nmol/mol for Eu, Tb, Lu, and Ym, also show coherent seasonal variability, along with total REEs (Fig. 2.7). High coastal seawater REE levels are usually delivered by terrestrial runoff during the rainy season [6, 35]. In contrast, the high WZI-1 coral REE levels are observed in the dry winter/early spring season (Figs. 2.7 and 2.S3).

WZI, a small island with an area of 25 km^2, is isolated from mainland China, well-protected with a dense vegetation cover and lack of riverine sediment input and local anthropogenic inputs. The most probable source for high REE values is from aeolian dust transported by winter monsoon winds. East Asia is a primary source for natural and anthropogenic dust aerosols in the Northern Hemisphere [38]. The strong and dry winter monsoon not only brings cold air masses but also transports large amounts of dust (average deposition: 2–5 $g/m^2/year$) to eastern and southern China and beyond [11, 37], recorded as high REE levels in WZI corals. This high-resolution coral REE fingerprints could be new proxies for East Asian winter monsoon.

2.4 Summary and Conclusions

After carefully addressing the issues of spectral interferences, blank and memory effects, ratio drifting and mass discrimination, and matrix effects, a precise technique for measuring carbonate REEs on SF-ICPMS has been developed. A high instrumental sample throughput of 8–10 samples per hour can be achieved routinely. This high-sensitivity method with good external reproducibility (1.9–6.5%) requires only 10–1000 fg of REEs in 10–20 μg of carbonate samples.

Compared with labor-intensive TIMS methods [17, 30], only simple cleaning and dissolution steps are involved and no time-consuming column separation is required in this new technique. The analytical precision is superior to that of previous ICPMS techniques (e.g., [4, 1]) and modern LA-ICPMS techniques (e.g., [6, 35]). A clear seasonal cycle of low WZI coral REE/Ca ratios <1 nmol/mol shown by our fast SF-ICPMS method is difficult to achieve using any other current ICPMS technique. The examples described here demonstrate the applicability of high-precision monthly coral REEs as proxies for detailed dust history and monsoon evolution. In addition to corals and foraminifera, our technique can also be applied to diverse carbonates, such as sclerosponges, tufa, shells, and speleothems. Hence, the carbonate REE-related studies of climate change [39], environmental research [6, 31] and paleoceanography [9] can benefit from the improvements in precision, sample size, and sample throughput inherent in this new SF-ICPMS method.

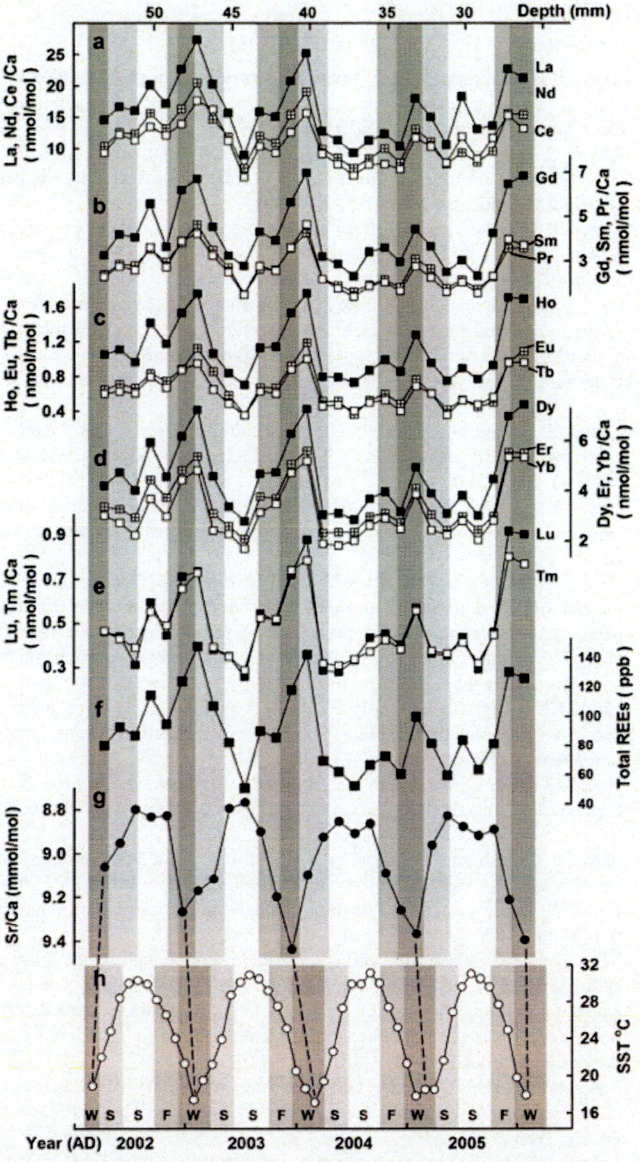

Fig. 2.7 Monthly resolved time series for the 4-year period (AD 2002–2005) of **a** La/Ca, Nd/Ca, and Ce/Ca; **b** Gd/Ca, Sm/Ca, and Pr/Ca; **c** Ho/Ca, Eu/Ca, and Tb/Ca; **d** Dy/Ca, Er/Ca, and Yb/Ca; **e** Lu/Ca and Tm/Ca; **f** total REEs; and **g** Sr/Ca [27] for the *Porites* coral, WZI-1, and **h** monthly sea surface temperature (SST) from a WZI meteorological station. Coral REE time series were calendared by aligning coral Sr/Ca maximums with SST minimums. The error bars of REE data are smaller than the symbol size. Reprinted from Ref. [29] with permission of copyright American Chemical Society

References

1. Akagi T, Hashimoto Y, Fu FF, Tsuno H, Tao H, Nakano Y (2004) Variation of the distribution coefficients of rare earth elements in modern coral-lattices: species and site dependencies1. Geochim Cosmochim Acta 68(10):2265–2273
2. Alibert C, Kinsley L, Fallon SJ, McCulloch MT, Berkelmans R, McAllister F (2003) Source of trace element variability in Great Barrier Reef corals affected by the Burdekin flood plumes. Geochim Cosmochim Acta 67(2):231–246
3. Byrne RH, Sholkovitz ER (1996) Marine chemistry and geochemistry of the lanthanides. In: Gschneidner Jr KA, Eyring L (eds) Handbook of the physics and chemistry of rare earths, vol 23, pp 497–593
4. Eggins SM, Woodhead JD, Kinsley LPJ, Mortimer GE, Sylvester P, McCulloch MT et al (1997) A simple method for the precise determination of ≥40 trace elements in geological samples by ICPMS using enriched isotope internal standardisation. Chem Geol 134(4):311–326
5. Elderfield H, Greaves MJ (1982) The rare earth elements in seawater. Nature 296(5854):214–219
6. Fallon SJ, White JC, McCulloch MT (2002) *Porites* corals as recorders of mining and environmental impacts: Misima Island, Papua New Guinea. Geochim Cosmochim Acta 66(1):45–62
7. Field MP, Sherrell RM (1998) Magnetic sector ICPMS with desolvating micronebulization: interference-free subpicogram determination of rare earth elements in natural samples. Anal Chem 70(21):4480–4486
8. Green DRH, Cooper MJ, German CR, Wilson PA (2003) Optimization of an inductively coupled plasma–optical emission spectrometry method for the rapid determination of high-precision Mg/Ca and Sr/Ca in foraminiferal calcite. Geochem Geophys Geosystems 4(6)
9. Haley BA, Klinkhammer GP, Mix AC (2005) Revisiting the rare earth elements in foraminiferal tests. Earth Planet Sci Lett 239(1–2):79–97
10. Jarvis KE, Gray AL, McCurdy E (1989) Avoidance of spectral interference on europium in inductively coupled plasma mass spectrometry by sensitive measurement of the doubly charged ion. J Anal At Spectrom 4(8):743–747
11. Jickells TD, An ZS, Andersen KK, Baker AR, Bergametti G, Brooks N et al (2005) Global iron connections between desert dust, ocean biogeochemistry, and climate. Science 308(5718):67–71
12. Kawabe I, Kitahara Y, Naito K (1991) Non-chondritic yttrium/holmium ratio and lanthanide tetrad effect observed in pre-Cenozoic limestones. Geochem J 25(1):31–44
13. Lea DW, Boyle EA (1991) Barium in planktonic foraminifera. Geochim Cosmochim Acta 55(11):3321–3331
14. McLennan SM (1989) Rare earth elements in sedimentary rocks; influence of provenance and sedimentary processes. Rev Mineral Geochem 21(1):169–200
15. Murphy KE, Long SE, Rearick MS, Ertas ÖS (2002) The accurate determination of potassium and calcium using isotope dilution inductively coupled "cold" plasma mass spectrometry. J Anal At Spectrom 17(5):469–477
16. Nozaki Y, Lerche D, Alibo DS, Snidvongs A (2000) The estuarine geochemistry of rare earth elements and indium in the Chao Phraya River Thailand. Geochim Cosmochim Acta 64(23):3983–3994
17. Palmer MR (1985) Rare earth elements in foraminifera tests. Earth Planet Sci Lett 73(2–4):285–298
18. Raut NM, Huang LS, Lin KC, Aggarwal SK (2005) Uncertainty propagation through correction methodology for the determination of rare earth elements by quadrupole based inductively coupled plasma mass spectrometry. Anal Chim Acta 530(1):91–103
19. Richter DK, Gotte T, Niggemann S, Wurth G (2004) REE3+ and Mn2+ activated cathodoluminescence in late glacial and Holocene stalagmites of central Europe: evidence for climatic processes? Holocene 14(5):759–768

20. Roberts NL, Piotrowski AM, McManus JF, Keigwin LD (2010) Synchronous deglacial overturning and water mass source changes. Science 327(5961):75–78
21. Rosenthal Y, Field MP, Sherrell RM (1999) Precise determination of element/calcium ratios in calcareous samples using sector field inductively coupled plasma mass spectrometry. Anal Chem 71(15):3248–3253
22. Scherer M, Seitz H (1980) Rare-earth element distribution in Holocene and Pleistocene corals and their redistribution during diagenesis. Chem Geol 28:279–289
23. Shaw HF, Wasserburg GJ (1985) Sm-Nd in marine carbonates and phosphates: implications for Nd isotopes in seawater and crustal ages. Geochim Cosmochim Acta 49(2):503–518
24. Shen CC, Lee T, Chen CY, Wang CH, Dai CF, Li LA (1996) The calibration of *D*[Sr/Ca] versus sea surface temperature relationship for *Porites* corals. Geochim Cosmochim Acta 60(20):3849–3858
25. Shen CC, Hastings DW, Lee T, Chiu CH, Lee MY, Wei KY, Edwards RL (2001) High precision glacial–interglacial benthic foraminiferal Sr/Ca records from the eastern equatorial Atlantic Ocean and Caribbean Sea. Earth Planet Sci Lett 190(3–4):197–209
26. Shen CC, Edwards RL, Cheng H, Dorale JA, Thomas RB, Moran SB et al (2002) Uranium and thorium isotopic and concentration measurements by magnetic sector inductively coupled plasma mass spectrometry. Chem Geol 185(3–4):165–178
27. Shen CC, Chiu HY, Chiang HW, Chu MF, Wei KY, Steinke S et al (2007) High precision measurements of Mg/Ca and Sr/Ca ratios in carbonates by cold plasma inductively coupled plasma quadrupole mass spectrometry. Chem Geol 236(3–4):339–349
28. Shen CC, Li KS, Sieh K, Natawidjaja D, Cheng H, Wang X et al (2008) Variation of initial $^{230}Th/^{232}Th$ and limits of high precision U-Th dating of shallow-water corals. Geochim Cosmochim Acta 72(17):4201–4223
29. Shen CC, Wu CC, Liu Y, Yu J, Chang CC, Lam DD et al (2011) Measurements of natural carbonate rare earth elements in femtogram quantities by inductive coupled plasma sector field mass spectrometry. Anal Chem 83(17):6842–6848
30. Sholkovitz E, Shen GT (1995) The incorporation of rare earth elements in modern coral. Geochim Cosmochim Acta 59(13):2749–2756
31. Siklósy Z, Demeny A, Vennemann TW, Pilet S, Kramers J, Leél-Őssy S, Shen C-C, Hegner E (2009) Bronze Age volcanic event recorded in stalagmites by combined isotope and trace element studies. Rapid Commun Mass Spectrom 23(6):801–808
32. Taylor SR, McLennan SM (1995) The geochemical evolution of the continental crust. Rev Geophys 33(2):241–265
33. Webb GE, Kamber BS (2000) Rare earth elements in Holocene reefal microbialites: a new shallow seawater proxy. Geochim Cosmochim Acta 64(9):1557–1565
34. Wollenweber D, Straßburg S, Wünsch G (1999) Determination of Li, Na, Mg, K, Ca and Fe with ICP-MS using cold plasma conditions. Fresenius' J Anal Chem 364(5):433–437
35. Wyndham T, McCulloch M, Fallon S, Alibert C (2004) High-resolution coral records of rare earth elements in coastal seawater: biogeochemical cycling and a new environmental proxy. Geochim Cosmochim Acta 68(9):2067–2080
36. Yu J, Day J, Greaves M, Elderfield H (2005) Determination of multiple element/calcium ratios in foraminiferal calcite by quadrupole ICP-MS. Geochemistry, Geophysics, Geosystems, 6(8): Q08P01; DOI:10.1029/2005GC000964.
37. Zhang J, Nozaki Y (1996) Rare earth elements and yttrium in seawater: ICP-MS determinations in the East Caroline, Coral Sea, and South Fiji basins of the western South Pacific Ocean. Geochim Cosmochim Acta 60(23):4631–4644
38. Zhang XY, Arimoto R, An ZS (1997) Dust emission from Chinese desert sources linked to variations in atmospheric circulation. J Geophys Res 102(D23):28041–28047
39. Zhou H, Wang Q, Zhao J, Zheng L, Guan H, Feng Y, Greig A (2008) Rare earth elements and yttrium in a stalagmite from Central China and potential paleoclimatic implications. Palaeogeogr Palaeoecol 270(1–2):128–138

Niño phase were primarily responsible, under current global warming conditions. The SCS, the largest marginal sea of the Pacific Ocean, is situated in a meteorologically sensitive zone. Its southern extremity is close to the Western Pacific Warm Pool, whose climate dynamics are intimately related to El Niño-Southern Oscillation (ENSO) variability [30]. However, summertime SSTs in the SCS [41] between 1991 and adjacent years (1987–1996) do not show any significant temperature differences (Figs. 3.S1 and 3.S2), indicating that other factors must have fueled the 1991 summer coral mortality event there. Here the compelling results of REE contents, trace element Al/Ca values, and skeletal textural features all responded to the 1991 Mount Pinatubo eruption.

3.2 Materials and Methods

3.2.1 Sample Collection

A live, massive, 35-cm Porites coral head, ST0506, from a water depth of 4 m, off the coast of Son Tra Island (STI, 16°13′ N, 108°12′ E), in central Vietnam and facing the western SCS (Figs. 3.1 and 3.S3a) was collected. An additional supplementary *Porites* coral core, MJL01a, was collected from a water depth of 3–4 m in Meiji Reef (MJR, 9°55′ N, 115°32′ E), central SCS, in an open-ocean environment surrounded by a reef flat with a large central lagoon (Figs. 3.1 and 3.S4a).

3.2.2 Geochemical Analysis

Continuous 4-year rare earth elements (REE) contents with monthly-to-bimonthly resolution from AD 1990 to 1993 for both corals were analyzed on a solution nebulization-inductively coupled plasmas mass spectrometer (SN-ICPMS), Thermo Electron ELEMENT II [34] (**Supplementary information**). Ba/Ca, Al/Ca, and Sr/Ca of coral ST0506 with clean treatments [33] were also measrued [20]. The micro-domain skeletal textures of three selected depths of coral ST0506 were analyzed by a field emission scanning electron microscope equipped with an energy dispersive X-ray spectrometer (FESEM/EDS) (**Supplementary information**).

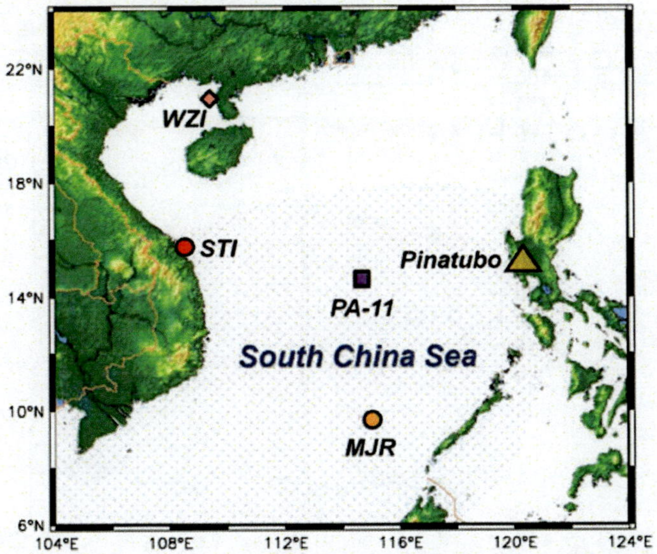

Fig. 3.1 Map of the South China Sea (SCS) region. Symbols denote locations of STI coral (red circle), MJR coral (orange circle), WZI coral (pink diamond), open SCS seawater (purple square, station PA-11) [1] and Mount Pinatubo dacite (yellow triangle). The shaded area represents the fallout distribution territory of the Pinatubo volcanic eruption in mid-June 1991 (modified from [28]). Reprinted from Wu et al. [50] with permission of copyright American Geophysical Union

3.3 Results

3.3.1 Temporal Variations of REE/Ca Ratios

Measured skeletal REE/Ca ratios for coral ST0506 are plotted in Fig. 3.S5. La/Ca and Yb/Ca ratios are highlighted as representative for the changes of the mass fractions of light rare earth elements (LREE, La-Eu) and heavy rare earth elements (HREE, Gd-Lu) to Ca (Fig. 3.2a, b). Distinct and coherent abrupt increases of coral skeletal REE/Ca contents can be observed from spring to summer 1991. Different element/Ca ratios at the peak display an anomaly ranging from 3.2–57 times higher than baseline values, from adjacent years in spring 1990 to early 1991 (Figs. 3.2a, b, and 3.S5). The sudden REE enrichment ended at a hiatus observed in mid-late 1991. After a 2–3 month discontinuity [35], the coral recovered in the early 1992, and the magnitude of REE/Ca values assumed baseline levels again (Figs. 3.2a, b, and 3.S5).

In coral ST0506, La/Ca values increased from 10–20 nmol/mol before 1991 spring and climbed to a peak of 678 nmol/mol in the summer of 1991, where the position is prior to the onset of the hiatus (Fig. 3.2a). This maximum is 30–40 times higher than values observed in other baseline years. For the segment above the hiatus, La/Ca values dropped to 15–50 nmol/mol in the early 1992. Yb/Ca ratios show a maximum

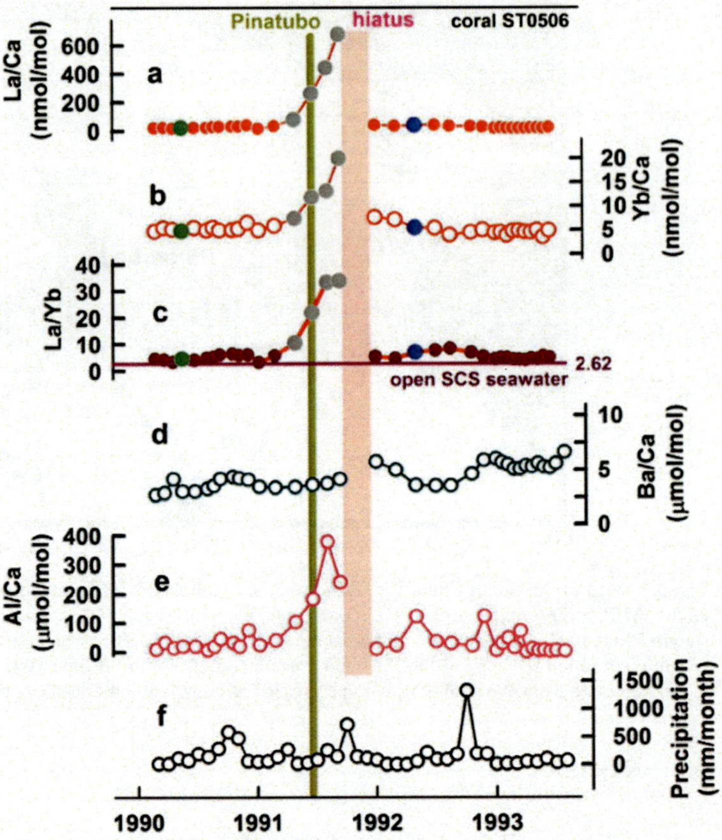

Fig. 3.2 Monthly-to-bimonthly-resolved coral ST0506 REE records between AD 1990 and 1993. **a** La/Ca ratios, **b** Yb/Ca ratios, **c** La/Yb ratios. Purple line shows La/Yb baseline values from an open SCS seawater station [1]. **d** Ba/Ca ratios, **e** Al/Ca ratios, and **f** monthly instrumental precipitation from 1990 to 1993 (from Da Nang Weather Station). Pink bar indicates the duration of a coral growth hiatus. Dark yellow line denotes the climactic eruption of Pinatubo on June 15, 1991. Green, grey, and blue circles represent the layers corresponding to the REE patterns shown in Fig. 3.3b. Reprinted from Wu et al. [50] with permission of copyright American Geophysical Union

of 20 nmol/mol in the summer of 1991, 4 times higher than values between 1990 and 1993 (Fig. 3.2b).

Similar REE/Ca trends and peaks during the same period in the summer of 1991 are registered in coral MJL01a (Fig. 3.S6). La/Ca and Yb/Ca ratios increased from 4.2 to 12 nmol/mol and 1.4–2.8 nmol/mol, respectively (Fig. 3.S6a, S6b), 2–3 times higher than central SCS baseline values.

3.3.2 Changes in Coral REE Diagrams

Representative coral ST0506 REE contents for pre-anomaly, anomaly, and post-anomaly, normalized to the post-Archean Australian shale (PAAS) [23], show a distinct interruption during 1990–1992 (Fig. 3-3b and Table 3.S1). The REE diagram in April 1990 exhibits an increasing trend from Ce to Lu and a negative Ce anomaly, as typical of baseline layers. It represents the characteristics of a "no-event" coral sample with a symbolic "seawater-like" REE pattern [37] (Fig. 3.3c), as observed in a coral from Weizhou Island (WZI, 21°01′ N, 109°04′ E) [34], northwest SCS (Fig. 3.1). From April to August, 1991, REE/Ca data display pronounced increases in LREEs and HREEs, and a gradual transformation of the negative Ce anomaly towards more a positive anomaly in the REE diagrams (grey records in Figs. 3.2a, b,

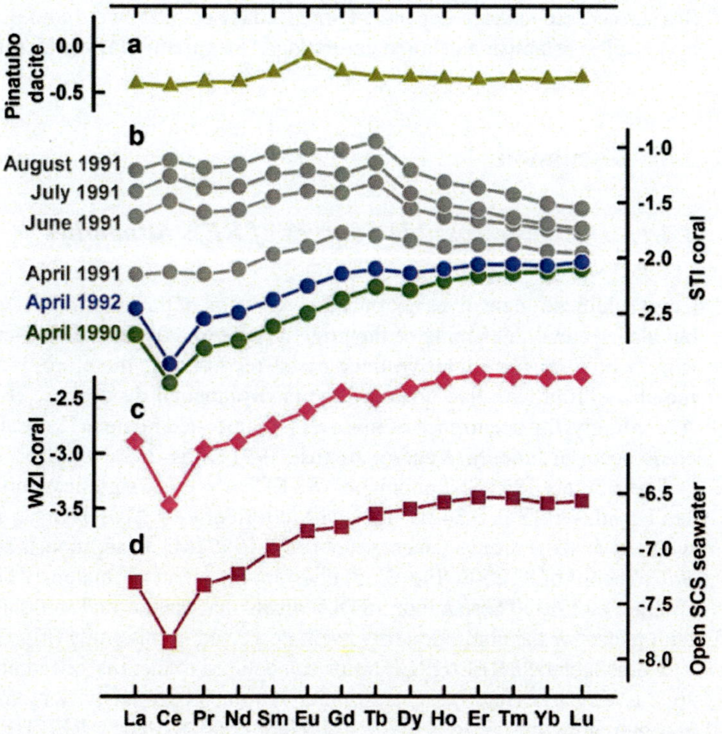

Fig. 3.3 Comparison of shale-normalized REE patterns. **a** Pinatubo dacite. **b** Coral ST0506 REE patterns for period before the 1991 Pinatubo eruption (April 1990, green circle), between-and-after the eruption (June to August 1991, grey circle), and post eruption (June 1992, blue circle). **c** WZI coral [34]. (d) Open SCS seawater [1]. Reprinted from Wu et al. [50] with permission of copyright American Geophysical Union

and 3.3b). The REE diagrams during June–August 1991 (Fig. 3.3b) with a LREE-enriched and HREE-depleted configuration demonstrate a significant discrepancy from the baseline signatures. These atypical features faded from late 1991 to early 1992. In April 1992, the REE diagram resumes a typical open-ocean seawater pattern with a negative Ce anomaly and HREE enrichment (blue records in Fig. 3.3b).

3.3.3 Ba/Ca, Al/Ca, and Sr/Ca Records of Coral ST0506

Coral ST0506 Ba/Ca and Al/Ca from 1990–1993 are presented in Fig. 3.2d,e. Seasonal Ba/Ca data vary from 2.8 to 6.5 μmol/mol during AD 1990–1993. A pronounced Al/Ca pulse is found in the summer 1991 with peak values of 378 μmol/mol, 5–50 times higher than the baseline values of 8.0–100 μmol/mol. Sea surface temperature-dependent Sr/Ca data (Fig. 3.S5) were combined with ^{230}Th ages [35] to establish a detailed chronology (**Supplementary information**).

3.4 Discussion

3.4.1 Decipher Possible Sources of REE Anomalies

Coral calcification involves not only the extension of the outermost skeletal material, but also skeletal thickening of the growth layer to a depth of 0.5–1-cm tissue layer (Fig. 5 of [3]). The inclusive trace metal records are, therefore, biased by a few months with growth that preceded the environmental disturbance (Fig. 1 of [44]). Accordingly, the occurrence of this event, manifested in the SCS coral REE records, could occur in June, July, and/or August, 1991 (Figs. 3.2 and 3.S6).

The summer 1991 SST anomalies (SSTA) showed no significant difference at Son Tra Island (−0.13 ± 0.34 °C, 1σ) and Meiji Reef (−0.32 ± 0.49 °C, 1σ), compared with seasonal means in adjacent years 1987–1996 [41]. In addition, the thermal proxy Sr/Ca record of ST0506 (Fig. 3.S5) does not reflect an anomalously high SST in the summer of 1991. These minor SSTA indicate that the thermal stress alone was not a fatal factor for the coral mortality event described in this study (Fig. 3.S2).

Coral skeletal LREE/HREE ratios can be used to identify different REE sources, such as terrestrial runoff discharge, precipitation, bioproductivity, monsoon evolution and human activities 16, 18, 25, 47]. Here coral ST0506 LREE/HREE ratios, for example, La/Yb ratios, show relatively low values of 3.2–6.3 in 1990, close to a typical low surface seawater value of 2.62 (Fig. 3.2c) in a SCS open-ocean site [1],station PA-11, Fig. 3.1). An abrupt La/Yb increase occurred in spring-to-summer 1991, starting with a ratio of 10 and peaking at 34 (Fig. 3.2c). Supplementary coral MJL01a also shows records of about 2–3 times higher La/Yb ratios in mid-1991 (Fig. 3.S6c), and the timing of its abrupt increase matches that observed in coral

ST0506 (Fig. 3.2c). Pronounced LREE/HREE increases in the summer of 1991 were observed in both corals from coastal site in central Vietnam (Fig. 3.S3a) and an open-ocean atoll in the central SCS (Fig. 3.S4a), indicating that local terrestrial runoff [18] or pollution [5] cannot be the cause for the REE anomalies.

Skeletal Ba/Ca ratios in near-shore corals reflect precipitation, terrestrial discharge, and flood events [22, 39]. Coral ST0506 Ba/Ca values of 4.1 μmol/mol in the summer of 1991, are compatible with values of 3.0–6.6 μmol/mol in the adjacent summers. The muted Ba/Ca (Fig. 3.2d) and absence of high precipitation in the summer of 1991 (Fig. 3.2f) suggest that the decoupled high REE contents and La/Yb ratios in coral ST0506 cannot be attributed to regional terrestrial discharge or sediment reworking. These geochemical-based findings are also in good agreement with typhoon track evidence that no coeval typhoons or severe storms swept both coastal central Vietnam and the central SCS in the timing of summer 1991 (Fig. 3.S2). Cyclone forcing can thus be excluded.

Aluminum is one of the characteristic and abundant elements in igneous rocks, with a mass fraction range of 10^3–10^4 μg/g [26]. Al has been used as a trace indicator for volcanic events recorded in carbonates [40, 49]. Laboratory experiments demonstrated that high amounts of aluminum can be released when tephra enters surface seawater [7]. Our coral ST0506 exhibits extremely high Al/Ca ratios up to 400 μmol/mol, comparing to the baseline of 10–100 μmol/mol, displays a distinct pulse in the summer 1991 (Fig. 3.2e). The concurrent increase of Al/Ca and REE contents and ratios in 1991 (Fig. 3.2a–c), can be explained by the same igneous source.

3.4.2 REE Anomalies Linked to the 1991 Pinatubo Eruption

The second largest volcanic eruption of the twentieth century, the 1991 Mount Pinatubo in the Philippines (15°14′ N, 120°35′ E), was coincident with the REE anomalies. The entire duration of these eruptions occurred from April 2 to September 2. The precursory phreatic eruption of April 2 was small with minor quantities of ash to 500–800 m in height [8]. On June 15, the climactic eruption yielded a maximum altitude of 35–40 km ash and spread southwestward, across the SCS and further, to the Bay of Bengal [13, 28] (shaded area in Fig. 3.1). Large and widespread fallout was collected in sediment traps around the entire SCS region [48]. Hence, this eruption could have been responsible for the regional SCS coral mortality that occurred there in the summer of 1991.

REE diagrams of Pinatubo igneous fallout feature a relative flat pattern with a positive Eu anomaly [29], Fig. 3.3a). The coral ST0506 REE signature following the Pinatubo eruption shows a gradual transformation from a typical "seawater-like" REE pattern to a peculiar LREE-enriched and HREE-depleted pattern, with a positive Ce anomaly (Fig. 3.3b). This could result from a mixture of tephra and seawater REEs that became incorporated into the coral skeleton [36]. The relatively removal of HREEs in mid-1991 can be explained by preferential settling in silts and/or sands, where HREE-enriched minerals are concentrated [38].

The coral REE anomalies in ST0506 and MJL01a are nearly-synchronous in mid-1991 (Figs. 3.2a–c, 3.S5, and 3.S6). The REE peak in ST0506, however, occurs in August (Figs. 3.2a–c, and 3.S5), ~ 2 months later than the peak in MJL01a (Fig. 3.S6). This slight discrepancy is most likely be related to the different geographical settings between two sites. In the months following the climactic eruption in June, REEs and Al contents recorded in coral ST0506 were elevated dramatically by direct fallout of Pinatubo ash and from runoff delivery from the catchment area of the Han River to Da Nang Bay (Fig. 3.S3a). While over the threshold of environmental tolerance, this coral died in late 1991. The coral recovered in early 1992, and resumed a seawater-like REE pattern (Fig. 3.3b). Between spring and fall 1992, a subdued influx of La/Yb with a range of 4.6–8.3 occurred (Fig. 3.2c), suggesting that residual Pinatubo fallout on land continued to affect the coastal zone by precipitation or storms in the following year.

In coral MJL01a, the La/Ca peak coincides with the timing of the Pinatubo climactic eruption in June (Fig. 3.S6a), as well as a secondary influx after the spring of 1992 (Fig. 3.S6a), indicating that the dominant REE source was direct fallout. The La/Ca maximum is 13 times lower than that found in the ST0506 record (Figures S6a and 2a), that is subject to dilution since it is an open-ocean site (Fig. 3.1). The La/Yb plateau observed in both June and the following 1–2 month(s) (Fig. 3.S6c) suggest that the signals lingered in the surface water for a couple of months following the climactic eruption of June 15. The environmental impact at the open-ocean site was presumably mitigated by dilution as well.

3.4.3 *Micro-domain Images*

The images of micro-domain skeletal structure of coral ST0506 in different transverse sections by FESEM/EDS techniques are given in Fig. 3.4. Selected slices from normal growth layers in 1990 April (Fig. 3.4a) and 1992 June (Fig. 3.4c) contain only fine

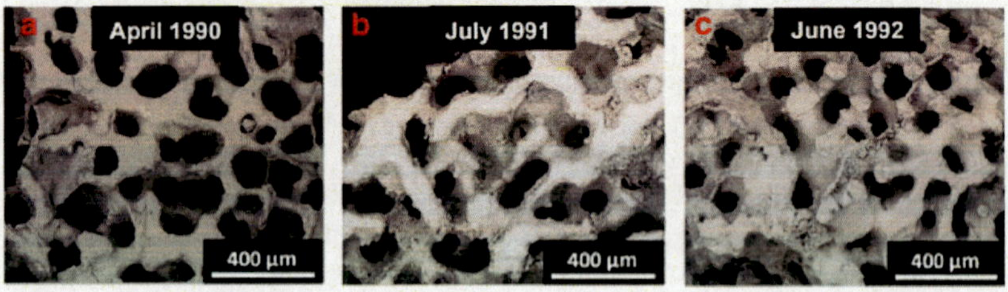

Fig. 3.4 SEM images of three subsamples of coral ST0506. Subsamples were sliced at depths of 15.2 cm, 14.1 cm, and 13.0 cm, corresponding to **a** April 1990, **b** July 1991, and **c** June 1992. Reprinted from Wu et al. [50] with permission of copyright American Geophysical Union

pristine skeletal carbonate. A slice from the 1991-July anomalous layer (Fig. 3.4b) after cleaning, shows that ash was incorporated within the voids of the coral lattice, with enriched matrices including not only C (16%), O (58%), and Ca (9.5%), but also significant amounts of Al (4.7%), Si (8.7%) and K (2.2%); as compared with baseline growth layers with only C (15%), O (50%), and Ca (35%). This micro-domain evidence further supports our interpretation of an ash source for the REEs. Hence, evidence from micro-domain features, REE and Al/Ca signatures (Figs. 3.4, 3.2a–c and 3.2e), all suggest that the elevated REEs and Al were at least partly incorporated from particulate phases.

3.4.4 The Connection Between Volcanism and Coral Mortality

When volcanic ash falls upon the ocean surface, it can release amounts of acids and metals to the environment, leading to acidification and surface water contamination [6]. Such factors are among the main contributors to reef degradation and coral mortality (e.g., [2, 52]). The image shown in Fig. 3.4b indicates that abundant ash covered the coral surface in the summer of 1991. Such an ash cover would directly interfere with polyp respiration, and decreasing photosynthesis by symbiotic zooanthellae. Ash suspended in the water column above would further stress the coral by reducing the amount of available light. Since coral reefs are becoming more and more fragile under the stress of global warming [15], corals would be especially vulnerable to an abrupt external event, such as the Pinatubo volcanic eruption.

3.5 Conclusions

This study demonstrates modern coral reefs were strongly exacerbated by a distant volcanic eruption, and yielding sensitive amplified responses under current high global thermal conditions. Contemporary continental near-shore reefs are more susceptible than their open-ocean counterparts. Given that the integrity of the biosphere has reached a high-risk planetary boundary in the Anthropocene [43], threats to reef vulnerability may lead to unprecedented biodiversity loss in the nearer future [15, 45]. Fully understanding the diverse biogeochemical processes and external impacts on coral reefs are needed for the sustainability of the biosphere.

References

1. Alibo DS, Nozaki Y (2000) Dissolved rare earth elements in the South China Sea: geochemical characterization of the water masses. J Geophys Res 105(C12):28771–28783
2. Anthony KR, Kline DI, Diaz-Pulido G, Dove S, Hoegh-Guldberg O (2008) Ocean acidification causes bleaching and productivity loss in coral reef builders. Proc Natl Acad Sci 105(45):17442–17446
3. Barnes DJ, Lough JM (1993) On the nature and causes of density banding in massive coral skeletons. J Exp Mar Biol Ecol 167(1):91–108
4. Brown BE, Dunne RP, Chansang H (1996) Coral bleaching relative to elevated seawater temperature in the Andaman Sea (Indian Ocean) over the last 50 years. Coral Reefs 15(3):151–152
5. Fallon SJ, White JC, McCulloch MT (2002) *Porites* corals as recorders of mining and environmental impacts: Misima Island, Papua New Guinea. Geochim Cosmochim Acta 66(1):45–62
6. Frogner P, Gislason SR, Óskarsson N (2001) Fertilizing potential of volcanic ash in ocean surface water. Geology 29(6):487–490
7. Frogner Kockum PC, Herbert RB, Gislason SR (2006) A diverse ecosystem response to volcanic aerosols. Chem Geol 231(1–2):57–66
8. Global Volcanism Program (1991) Report on Pinatubo (Philippines). In McClelland L (ed) Bulletin of the global volcanism network, vol 16, no 4. Smithsonian Institution. Retrieved from https://doi.org/10.5479/si.GVP.BGVN199104-273083
9. Glynn PW (1983) Extensive 'Bleaching' and death of reef corals on the pacific coast of Panamá. Environ Conserv 10(2):149–154
10. Glynn PW (1984) Widespread coral mortality and the 1982–83 El Niño warming event. Environ Conserv 11(2):133–146
11. Glynn PW (1996) Coral reef bleaching: facts, hypotheses and implications. Glob Change Biol 2(6):495–509
12. Hoegh-Guldberg O, Salvat B (1995) Periodic mass-bleaching and elevated sea temperatures: bleaching of outer reef slope communities in Moorea, French Polynesia. Mar Ecol Prog Ser 121:181–190
13. Holasek RE, Self S, Woods AW (1996) Satellite observations and interpretation of the 1991 Mount Pinatubo eruption plumes. J Geophys Res 101(B12):27635–27655
14. Hughes TP, Kerry JT, Álvarez-Noriega M, Álvarez-Romero JG, Anderson KD, Baird AH et al (2017) Global warming and recurrent mass bleaching of corals. Nature 543(7645):373
15. Hughes TP, Anderson KD, Connolly SR, Heron SF, Kerry JT, Lough JM et al (2018) Spatial and temporal patterns of mass bleaching of corals in the Anthropocene. Science 359(6371):80–83
16. Jiang W, Yu K, Song Y, Zhao J-X, Feng Y-X, Wang Y et al (2018) Annual REE signal of East Asian winter monsoon in surface seawater in the northern South China Sea: evidence from a century-long *Porites* coral record. Paleoceanogr Paleoclimatol 33: 168-178
17. Jiménez C, Cortés J (2002) Effects of the 1991–1992 El Niño on scleractinian corals of the Costa Rican central Pacific coast. Rev Biol Trop 49:239–250
18. Jupiter SD (2008) Coral rare earth element tracers of terrestrial exposure in nearshore corals of the Great Barrier Reef. Proc 11th Int Coral Reef Symp 1:105–109
19. Knowlton N, Brainard RE, Fisher R, Moews M, Plaisance L, Caley MJ (2010) Coral reef biodiversity. Life in the world's oceans: diversity distribution and abundance, pp 65–74
20. Lo L, Shen CC, Lu CJ, Chen YC, Chang CC, Wei KY et al (2014) Determination of element/Ca ratios in foraminifera and corals using cold-and hot-plasma techniques in inductively coupled plasma sector field mass spectrometry. J Asian Earth Sci 81:115–122
21. Lough JM, van Oppen MJH (2009) Introduction: coral bleaching—patterns, processes, causes and consequences. In: Coral bleaching. Springer, Berlin, Heidelberg, pp 1–5
22. McCulloch M, Fallon S, Wyndham T, Hendy E, Lough J, Barnes D (2003) Coral record of increased sediment flux to the inner Great Barrier Reef since European settlement. Nature 421(6924):727–730

23. McLennan SM (1989) Rare earth elements in sedimentary rocks; influence of provenance and sedimentary processes. Rev Mineral Geochem 21(1):169–200
24. Moberg F, Folke C (1999) Ecological goods and services of coral reef ecosystems. Ecol Econ 29(2):215–233
25. Nguyen AD, Zhao JX, Feng YX, Hu WP, Yu KF, Gasparon M et al (2013) Impact of recent coastal development and human activities on Nha Trang Bay, Vietnam: evidence from a *Porites* lutea geochemical record. Coral Reefs 32(1):181–193
26. Nockolds SR (1954) Average chemical compositions of some igneous rocks. Geol Soc Am Bull 65(10):1007–1032
27. Nyström M, Folke C, Moberg F (2000) Coral reef disturbance and resilience in a human-dominated environment. Trends Ecol Evol 15(10):413–417
28. Paladio-Melosantos MLO, Solidum RU, Scott WE, Quiambao RB, Umbal JV, Rodolfo KS et al (1996) Tephra falls of the 1991 eruptions of Mount Pinatubo. In: Newhall CG, Punongbayan R (eds) Fire and mud: eruptions and lahars of Mount Pinatubo, Philippines. Philippine Institute of Volcanology and Seismology, Quezon City, pp 513–535
29. Pallister JS, Hoblitt RP, Meeker GP, Knight RJ, Siems DF (1996) Magma mixing at Mount Pinatubo: petrographic and chemical evidence from the 1991 deposits. In: Newhall CG, Punongbayan R (eds) Fire and mud: eruptions and lahars of Mount Pinatubo, Philippines. Philippine Institute of Volcanology and Seismology, Quezon City, pp 687–731
30. Picaut J, Ioualalen M, Menkès C, Delcroix T, Mcphaden MJ (1996) Mechanism of the zonal displacements of the Pacific warm pool: Implications for ENSO. Science 274(5292):1486–1489
31. Reaka-Kudla ML (1997) The global biodiversity of coral reefs: a comparison with rain forests. Biodiversity II: Understanding and protecting our biological resources. Joseph Henry Press, Washington, DC, pp 83–108
32. Roberts CM, McClean CJ, Veron JE, Hawkins JP, Allen GR, McAllister DE et al (2002) Marine biodiversity hotspots and conservation priorities for tropical reefs. Science 295(5558):1280–1284
33. Shen C-C, Lee T, Liu K-K, Hsu H-H, Edwards RL, Wang C-H et al (2005) An evaluation of quantitative reconstruction of past precipitation records using coral skeletal Sr/Ca and $\delta^{18}O$ data. Earth Planet Sci Lett 237(3–4):370–386
34. Shen C-C, Wu C-C, Liu Y, Yu J, Chang C-C, Lam DD et al (2011) Measurements of natural carbonate rare earth elements in femtogram quantities by inductive coupled plasma sector field mass spectrometry. Anal Chem 83(17):6842–6848
35. Shen C-C, Wu C-C, Cheng H, Edwards RL, Hsieh Y-T, Gallet S et al (2012) High-precision and high-resolution carbonate ^{230}Th dating by MC-ICP-MS with SEM protocols. Geochim Cosmochim Acta 99:71–86
36. Sholkovitz ER, Landing WM, Lewis BL (1994) Ocean particle chemistry: the fractionation of rare earth elements between suspended particles and seawater. Geochim Cosmochim Acta 58(6):1567–1579
37. Sholkovitz E, Shen GT (1995) The incorporation of rare earth elements in modern coral. Geochim Cosmochim Acta 59(13):2749–2756
38. Sholkovitz ER, Elderfield H, Szymczak R, Casey K (1999) Island weathering: river sources of rare earth elements to the Western Pacific Ocean. Mar Chem 68(1–2):39–57
39. Sinclair DJ, McCulloch MT (2004) Corals record low mobile barium concentrations in the Burdekin River during the 1974 flood: evidence for limited Ba supply to rivers? Palerogeography, Palaeoclimatology, Palaeoecology 214(1):155–174
40. Siklósy Z, Demény A, Vennemann TW, Pilet S, Kramers J, Leél-Össsy S et al (2009) Bronze Age volcanic event recorded in stalagmites by combined isotope and trace element studies. Rapid Commun Mass Spectrom 23:801–808
41. Smith TM, Reynolds RW, Peterson TC, Lawrimore J (2008) Improvements to NOAA's historical merged land–ocean surface temperature analysis (1880–2006). J Clim 21(10):2283–2296
42. Spalding MD, Brown BE (2015) Warm-water coral reefs and climate change. Science 350(6262):769–771

43. Steffen W, Richardson K, Rockström J, Cornell SE, Fetzer I, Bennett EM et al (2015) Planetary boundaries: guiding human development on a changing planet. Science 347(6223):1259855
44. Taylor RB, Barnes DJ, Lough JM (1995) On the inclusion of trace materials into massive coral skeletons. 1. Materials occurring in the environment in short pulses. J Exp Marine Biol Ecology, 185(2):255–278
45. van Hooidonk R, Maynard JA, Planes S (2013) Temporary refugia for coral reefs in a warming world. Nat Clim Chang 3(5):508–511
46. Veron JEN (1995) Corals in space and time: the biogeography and evolution of the Scleractinia. Cornell University Press
47. Wyndham T, McCulloch M, Fallon S, Alibert C (2004) High-resolution coral records of rare earth elements in coastal seawater: biogeochemical cycling and a new environmental proxy. Geochim Cosmochim Acta 68(9):2067–2080
48. Wiesner MG, Wetzel A, Catane SG, Listanco EL, Mirabueno HT (2004) Grain size, areal thickness distribution and controls on sedimentation of the 1991 Mount Pinatubo tephra layer in the South China Sea. Bull Volcanol 66(3):226–242
49. Wu C-C, Burger M, Günther D, Shen C-C, Hattendorf B (2018) Highly-sensitive open-cell LA-ICPMS approaches for the quantification of rare earth elements in natural carbonates at parts-per-billion levels. Anal Chim Acta 1018:54–61
50. Wu C-C, Shen C-C, Lo L, Hsin Y-C, Yu K, Chang C-C et al (2018) Pinatubo volcanic eruption exacerbated an abrupt coral mortality event in 1991 summer. Geophys Res Lett 45:12396–12402
51. Yu KF, Zhao JX, Shi Q, Chen TG, Wang PX, Collerson KD et al (2006) U-series dating of dead Porites corals in the South China Sea: evidence for episodic coral mortality over the past two centuries. Quat Geochronol 1(2):129–141
52. Yu KF, Zhao JX, Collerson KD, Shi Q, Chen TG, Wang PX et al (2004) Storm cycles in the last millennium recorded in Yongshu Reef, southern South China Sea. Palaeogeogr Palaeoclimatol Palaeoecol 210(1):89–100

Chapter 4
Approach II: Highly-Sensitive Open-Cell Laser Ablation-Inductively Coupled Plasma Mass Spectrometry (LA-ICPMS) Approaches for the Quantification of Rare Earth Elements in Natural Carbonates at Parts-Per-Billion Levels

4.1 Introduction

The accurate determination of preserved high-resolution trace-element- and stable-isotope-variations in slowly growing natural carbonates, such as stalagmites, has recently gained extensive attention in the field of geology, earth- and environmental-sciences [6, 10, 12, 40, 42]. Stalagmites often grow continuously and can be accurately dated by means of radiometric techniques. Furthermore, they suffer little from secondary alteration and are therefore considered to be ideal multi-proxy archives for past climatic and environmental changes [8, 9, 28, 41, 44].

The geochemistry of rare earth elements (REEs) in speleothems, such as stalagmites, has been studied for understanding contemporary and late-quaternary climatic processes [19, 30, 47]. Since REEs are enriched in igneous rocks but depleted in carbonates [39], REE series preserved in stalagmites are considered to be an ideal indictor to reconstruct past gigantic volcano eruptions [35].

Diverse analytical methods, including neutron activation analysis (NAA) [31], inductively coupled plasma atomic emission spectrometry (ICPAES) [23], thermal ionization mass spectrometry (TIMS) [34], and inductively coupled plasma mass spectrometry (ICPMS) [43] have been employed for quantitative analyses of REEs in natural carbonates. Specifically, the detection capabilities offered by solution nebulization ICPMS allow multi-element analysis of carbonate REEs at ng/g to sub-ng/g levels [1, 2, 43]. However, these techniques were all based on discrete mini-domain sampling strategies and time-consuming sample preparation protocols, which prohibited acquisition of high-resolution records (<500 μm) [37] and limited sample throughput. In 2011, we presented a fast and precise method, using

Electronic supplementary material The online version of this chapter (https://doi.org/10.1007/978-981-16-3619-6_4) contains supplementary material, which is available to authorized users.

C.-C. Wu, *Advanced and Applied Studies on Ultra-Trace Rare Earth Elements (REEs) in Carbonates Using SN-ICPMS and LA-ICPMS*, Springer Theses,
https://doi.org/10.1007/978-981-16-3619-6_4

sector field-inductively coupled plasma mass spectrometry (SF-ICPMS) for the determination of carbonate REEs in femtogram quantities without chemical separation processes (details described in the Chap. 3). Although a high sample throughput of 8–10 samples per hour could be achieved, steps of physical subsampling, dissolution, and dilution were still required [32].

Laser ablation sampling in combination with inductively coupled plasma mass spectrometry (LA-ICPMS) offers the capabilities for the rapid, in-situ determination of major, minor and trace element concentrations [16, 18]. High sensitivity, large linear dynamic range and the possibility to collect data at high lateral resolution between single digit and hundreds of μm make LA-ICPMS a powerful tool for element- and/or isotope-specific microanalyses in solid materials [25].

In conventional LA-ICPMS analyses, the sample of interest needs to be placed inside an enclosing ablation chamber that limits the size of objects that can be analyzed to several centimeters. This constrains the applicability of LA-ICPMS for the analysis of large-scale stalagmites. In 2013, Tabersky et al. introduced an open-cell LA-ICPMS approach that allows analysis of differently sized and shaped samples in an atmospheric environment. Large samples do not need to be cut to fit within an ablation cell anymore, which prevents sample damage and avoids loss of information. Validity of this approach has been tested with various standard reference materials (SRMs). Suitability for the analysis of large-scale stalagmite samples has been demonstrated. REE concentrations in the several hundreds of ng/g regime could be quantified in glass SRMs using the open-cell LA approach in combination with an inductively coupled plasma quadrupole mass spectrometer (Q-ICPMS). However, this setup was not sensitive enough to determine REE profiles in the stalagmite sample. This was due to the very low REE abundances in natural carbonates, typically ranging from several hundreds of pg/g to several tens of ng/g [4, 47].

Here an advanced LA-ICPMS methodology based on the combination of open-cell LA with a SF-ICPMS was presented. The use of a jet-interface with the addition of small amounts of N_2 gas allowed a 20–40 fold sensitivity enhancement, which translated into a more than one order of magnitude improvement in limits of detection (LODs). This enabled the quantitative determination of speleothem REE concentrations in the single digit ng/g range. Emphases were placed on the investigation of mass load effects [26], the evaluation of quantification capabilities, and detection power. Following careful method development, the open-cell LA-SF-ICPMS approach was applied to a stalagmite collected from East Timor. REE concentration profiles showed several possible volcano events within the time periods of 79–86 thousand years ago (ka, before 1950 AD, hereafter).

4.2 Analytical Procedure

4.2.1 Instrumentation

All experiments were carried out using a 193 nm ArF excimer LA system (GeoLas C, Lambda Physik, Göttingen, Germany) coupled to a SF-ICPMS (Element XR, Thermo Fisher Scientific, Bremen, Germany). LA was performed on standard reference materials (SRMs) with either a conventional closed ablation chamber, or an open-cell setup. ICPMS detection was carried out running the Element XR instrument with either standard- or jet-interface configuration.

The closed-cell setup consisted of an air-tightly sealing cylindrical ablation chamber that could hold samples of up to 50 mm diameter and offered a total volume of 63 mm^3 [17]. The open-cell setup comprised an aerosol entrainment device (MANTA) that allowed LA sampling from almost arbitrarily sized objects and a large-capacity gas exchange device (GED, J-SCIENCE Lab Co., Ltd., Kyoto, Japan). By means of a membrane pump, the laser-generated aerosols were extracted from the sampling site and delivered to the GED. Within the GED, the air was exchanged with argon to allow for operation of the ICP [36]. Specifically, with the open-cell setup, LA sampling was just limited by the scanning range of the translation stage (>230 mm in length, 34 mm in width, and 16 mm in height). Figure 4.1 provides a schematic representation of the open-cell LA-SF-ICPMS approach. A detailed description and evaluation of this methodology can be found in the literature [36].

In standard configuration, the vacuum interface of the SF-ICPMS comprised a standard sampler cone and a nickel X skimmer cone (X-version, Thermo Fisher Scientific, Waltham, USA). In jet configuration, the standard sampler cone was replaced with a jet sampler cone (JET-version, Thermo Fisher Scientific, Waltham, USA). Additionally, a high capacity vacuum pump was used to maintain the pressure gradient between the ICP, the interface region and the mass analyzer. Small amounts of N_2 gas were admixed to the carrier gas stream before the aerosol was introduced into the ICP torch.

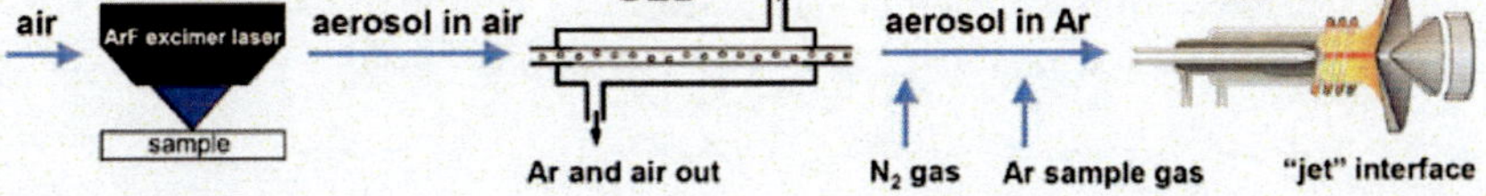

Fig. 4.1 Schematic representation of the open-cell LA-SF-ICPMS approach. The airborne aerosol is introduced into a GED where the air is exchanged with argon. Upstream of the ICP torch, Ar carrier gas is added. When the jet-interface is used, a minute amount of N_2 is supplied as well. Detection is carried out by SF-ICPMS. Reprinted from Wu et al. [45] with permission of copyright Elsevier

Table 4.1 Operating parameters of the LA-SF-ICPMS setup

Laser ablation system: GeoLas C	
Wavelength	193 nm
Pulse duration	16 ns
Laser fluence	14–17 J/cm^2
Laser spot size	16–160 μm
Laser repetition rate	5–40 Hz
*Carrier gas flow (He)	1.0–1.2 L/min
#Carrier gas flow (Air)	0.8–0.9 L/min
#Sweep gas flow (Ar)	8.0–10 L/min
ICPMS: Element XR	
Sample gas flow (Ar)	0.8–1.0 L/min
Auxiliary gas flow (Ar)	0.8–1.2 L/min
Cooling gas flow (Ar)	16 L/min
RF power	1225–1500 W
N_2 gas flow (jet-interface)	15–20 mL/min
Mass resolving power ($M/\Delta M$)	Low (300)
Samples per peak	20
Integration time	8 ms
Integration window	40%
Integration type	Average
Time per sweep	0.49 s
Distance per sweep	4.9 μm
Detector mode	Triple
$^{232}Th^{16}O^+/^{232}Th^+$	0.3–0.5%
$^{238}U^+/^{232}Th^+$	0.98–1.02

*applied only to the closed-cell setup
#applied only to open-cell setup

Operating parameters of the LA-SF-ICPMS setup are reported in Table 4.1. They were adjusted to allow maximum sensitivity for ^{238}U while maintaining high sensitivity for a broad mass to charge (*m/z*) range, low oxide formation rates and similar ^{238}U and ^{232}Th sensitivities. They were determined for every instrumental approach separately from ablation of NIST SRM 612 with a 16 μm diameter circular laser spot, 10 Hz ablation frequency and 10 μm/s scan speed.

4.2.2 Experimental Design

Sensitivities have been determined for the closed-cell and the open-cell setup. Both setups have been tested with the standard- and the jet-interface. LA was carried out on NIST SRM 612 with laser spots of 160 μm, 10 Hz ablation frequency and a scan speed of 10 μm/s.

Mass load effects were investigated by evaluating the abundance corrected sensitivities and the variation of sensitivity ratios to the internal standard as a function of the area ablated with circular laser spots of variable diameters in the range from 16 μm to 160 μm. Given the homogeneous beam profile delivered by the LA system used in this study, the area ablated with a certain spot is considered a direct measure for the mass load introduced into the system.

Quantification capabilities of the open-cell LA-SF-ICPMS approach were tested in analyses of NIST SRM 612, NIST SRM 614 and USGS MACS-3. Relative sensitivity factors were determined from ablation of NIST SRM 610 (analysis of NIST SRM 612 and NIST SRM 614) and NIST SRM 614 (analysis of USGS MACS-3).

4.2.3 Data Collection, Processing, and Evaluation

Data were acquired following a standard sample standard bracketing protocol comprising data from acquisitions of an external reference material to establish relative sensitivities, unknown samples, and an external reference material again to correct for instrument drift. This sequence is a conventional protocol for ICPMS analyses such as Q-ICMPS [35, 36] or time-of-flight ICPMS (TOF-ICPMS) [3, 5]. For each individual acquisition, gas blank signal intensities were determined and used as a measure for background signal intensities and subtracted from the signals recorded during ablation.

Quantification was performed using an in-house developed software package STALQUANT [13] that follows the algorithm proposed by Longerich et al. [27]. Two different quantification schemes were used for calculation of high-resolution concentration profiles and determination of bulk element abundances. High-resolution concentration profiles were established by relating the mass scan acquisition time to the LA sampling position. In these experiments, every data point of a given time trace was quantified individually. Unlike that, bulk concentrations were determined from isotope intensities averaged over the ablation period. In all experiments, ^{44}Ca was selected as the internal standard isotope.

The critical level of an analysis has commonly been calculated by taking three times the standard deviation of a gas blank as a measure for the background noise [27]. Here, the instrumental REE background intensities consisted of stochastic single ion counts. Thus, according to the recommendation of the International Union of Pure and Applied Chemistry (IUPAC), LODs were calculated assuming counting statistics [7] as outlined in the supporting information.

4.2.4 Stalagmite Sample

The stalagmite sample MC 4–2 was collected from Lekiraka cave in East Timor (8°47′10.8″S, 126°23′31.1″E). The cave is located in a region surrounded by hundreds of volcanoes, which form a part of the "Pacific Ring of Fire". The stalagmite sample was sliced into slabs of 1 cm thickness and their surfaces were polished. Slab MC 4-2 analyzed in this study is 15 cm in length and 5 cm in diameter. From slab MC 4-2, several drilling subsamples of 50–100 mg each were taken for U-Th dating [33]. U-Th ages indicated that this stalagmite grew in the time period from 101 to 73 ka.

The stalagmite sample MC 4-2 was analyzed using the open-cell LA-SF-ICPMS approach developed in this study. Spatially resolved trace element concentration profiles were calculated from time traces recorded in two individual line scanning experiments. In both studies, the stalagmite was scanned across a distance of 8 mm that covered the time periods from 88 to 78 ka. The ablation tracks were running in parallel to each other and were separated by several 100 μm. Both experiments were carried out with a 160 μm diameter circular laser spot and an ablation frequency of 10 Hz. The scan speed of 10 μm/s resulted in a lateral displacement of 4.9 μm for an entire mass scan. Prior to the acquisition of high-resolution concentration profiles, the sample surface was cleaned with a low energy LA routine (3.7 J/cm fluence, 160 μm diameter circular laser spot, 10 Hz ablation frequency, 10 μm/s scan speed).

The average elemental composition of the anomaly layer and the adjacent baseline regions were investigated with hole drilling experiments. These studies were performed with a 160 μm diameter circular laser spot and 10 Hz ablation frequency.

For line scanning and hole drilling experiments, NIST SRM 614 was used as the calibration standard to establish relative sensitivity factors. Specifically, NIST SRM 614 was selected as external reference material [11, 29, 46] because its lower REE levels (720–813 ng/g) better match the expected speleothem REE concentrations than USGS MACS-3 or the more frequently used NIST SRM 612 or NIST SRM 610 [15, 35, 38].

Stoichiometry of the stalagmite sample was assumed as pure calcite and a Ca concentration of 400,000 mg/kg was used for internal standardization. This estimation cannot account for changes in the matrix composition and quantitative results will be biased when carbonate samples with variable concentrations of trace elements, organic material, and/or detritus are analyzed. However, this simplification was considered appropriate for the purpose of this study.

Signal intensities recorded from ablation of the stalagmite sample were occasionally spiking. This observation was attributed to remobilization of large particulates that got previously deposited at the walls of the MANTA, the aerosol transfer lines or the membranes within the GED. These spikes were removed from the time traces in an offline data processing routine. Data points that showed deviations exceeding a threshold of three times the standard deviation of the mean ion signal intensity calculated from its two adjacent data points were excluded from analysis. The average isotopic intensity was calculated as geometric mean, the standard deviation was derived assuming counting statistics.

4.3 Results and Discussion

4.3.1 Abundance Corrected Sensitivities as a Function of the Instrumental Setup

Figure 4.2 displays abundance corrected sensitivities for selected elements. Specifically, the isotope selection is covering a large proportion of the elemental *m/z* range and includes the complete suite of the REEs.

The use of the jet-interface in combination with the addition of small amounts of N_2 gas allowed a sensitivity enhancement by a factor of 20–40. This effect was observed for both, the closed-cell LA and the open-cell LA setup. For example, we found that the abundance corrected sensitivities for La were increased from 70 to

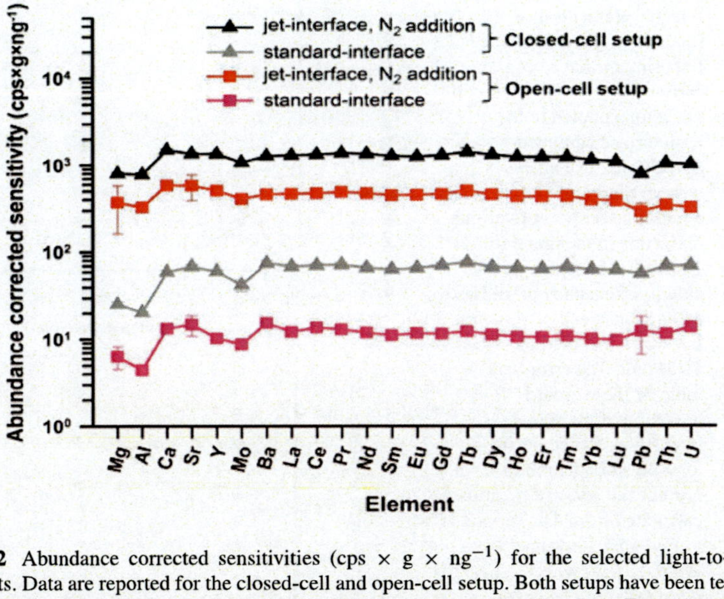

Fig. 4.2 Abundance corrected sensitivities (cps × g × ng^{-1}) for the selected light-to-heavy elements. Data are reported for the closed-cell and open-cell setup. Both setups have been tested in combination with the jet vacuum interface and the standard vacuum interface. Experiments were carried out on NIST SRM 612 with a 160 µm diameter circular laser spot, 10 Hz ablation frequency and 10 µm/s scan speed. Error bars indicate the standard deviation of the signal intensity recorded during at least 30 s of ablation. Reprinted from Wu et al. [45] with permission of copyright Elsevier

1300 cps × g × ng^{-1} (closed-cell LA) and from 13 to 460 cps × g × ng^{-1} (open-cell LA), respectively. This observation could be attributed to an improvement in ion extraction efficiency from the ICP and the first expansion stage enabled by the specific design of the jet sampler cone and the increased pumping speed [20, 24]. A 2-fourfold lower sensitivity observable with the open-cell LA setup has already been reported in a previously study [36]. This observation is explained with differences in transport efficiencies and material re-deposition related to the utilization of different carrier gases. For both LA setups, REE background signals intensities were maintained at similar values in the range from 0 to 1000 counts per second (cps) (Fig. 4.S1).

4.3.2 *Mass Load Effects*

Figure 4.3a shows abundance corrected sensitivities as a function of the area ablated with circular laser spots of various diameters (16, 24, 44, 60, 90, 120, 160 μm).

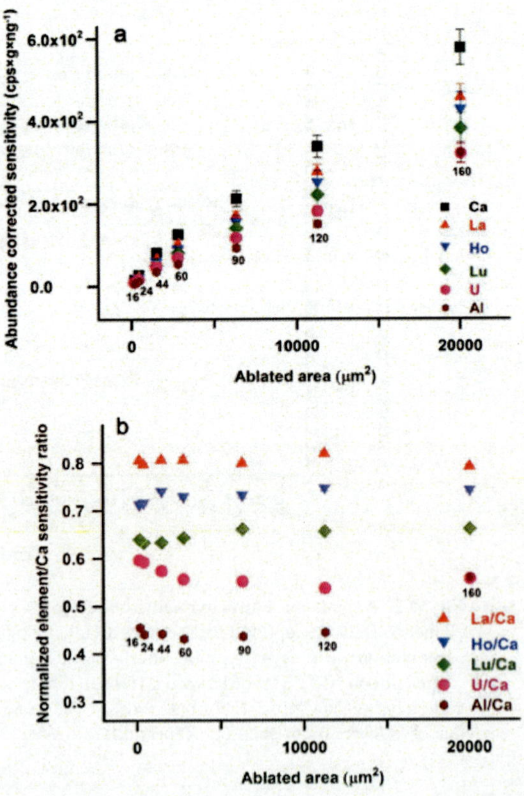

Fig. 4.3 **a** Abundance corrected sensitivities for the selected elements as a function of the area ablated. Experiments have been carried out with the open-cell LA setup coupled to the jet-interface equipped SF-ICPMS. The analyses were carried out with circular laser spots of various diameters in the range from 16 μm to 160 μm, constant ablation frequency of 10 Hz, and a scan speed of 10 μm/s. LA was performed on NIST SRM 612. The error bars indicate the standard deviation of the signal intensity recorded during at least 30 s of ablation. **b** Ca normalized sensitivity ratios calculated from data reported in Fig. 4.3a. Reprinted from Wu et al. [45] with permission of copyright Elsevier

Results are reported for Al, Ca, La, Ho, Lu and U. These elements were specifically selected to represent trace elements, matrix elements, light rare earth elements (LREE), middle rare earth elements (MREE), heavy rare earth elements (HREE) and actinide elements. Figure 4.3b displays the same data normalized to Ca.

The abundance corrected sensitivities reported in Fig. 4.3a did not increase proportionally to the area ablated with laser spots of variable diameters between 16 μm and 160 μm. For example, for a 100 fold increase in laser spot area, we determined an only 35 fold increase in abundance corrected sensitivity for La. The abundance corrected La sensitivities were 13 and 460 cps $\times$ g $\times$ ng^{-1} from 16 μm and 160 μm laser spots, respectively. Explanations for this observation include a reduced aerosol transport efficiency with increasing mass load. This could be a result of settling losses that became more significant with increasing mass ablation rate. Greater losses most likely occur within the MANTA because changing the laser repetition rate for a fixed spot size did follow a near-linear trend (Fig. 4.S2). The non-linear dependency of abundance corrected sensitivities from laser spot area is thus less likely a consequence of the changes in absolute aerosol density within the carrier gas, but rather related to the wider aerosol plume expansion after ablation with a larger laser spot.

Importantly, the sensitivity ratios of REEs relative to Ca remained nearly unaffected by the mass load introduced into the system (Fig. 4.3b). Relative standard deviations computed from seven measurements with spot sizes between 16 μm and 160 μm were 1.1%, 1.8% and 1.9% for La/Ca, Ho/Ca and Lu/Ca sensitivity ratios, respectively. These results are in agreement with previously reported findings [14, 22, 26] and indicate that there was little mass-load induced fractionation between REEs and Ca. However, Ca normalized U sensitivity ratios decreased by about 10% when the laser spot diameter was changed from 16 μm to 60 μm. This behavior is not yet fully understood but is not expected to have a strong influence on quantitative results obtained when mass load matching between external reference material and sample is provided. The significant difference in Ca normalized Al sensitivity ratios obtained from ablation with a 120 μm and a 160 μm laser spot is most likely an artifact caused by incorrect cross calibration between the analog mode of the secondary electron multiplier (SEM) which is used to detect Al signal intensity in experiments carried out with laser spots <120 μm and the Faraday detector, which was used to register Al signal intensities from 160 μm spots. However, trace element concentrations are orders of magnitude lower in stalagmites than they are in NIST SRM 612. They can be detected using the SEM only and issues related to an incorrect Faraday cross calibration are not expected to affect quantitative REE concentration profiles.

Based on these results, a 160 μm diameter circular laser spot, an ablation frequency of 10 Hz, and a scan speed of 10 μm/s were used for further experiments.

4.3.3 *Limits of Detection as a Function of the Instrumental Setup*

Figure 4.4 reports LODs calculated using Poisson-based statistics from ablation of NIST SRM 612 and the stalagmite sample MC 4-2 with a 160 µm diameter laser spot using 10 Hz ablation frequency and a scan speed of 10 µm/s.

For both, the closed-cell LA and the open-cell LA approach, the use of the jet-interface in combination with the addition of small amounts of N_2 gas allowed an improvement of LODs by one to two orders of magnitude. From integration of signal intensities recorded during 30 s of ablation (300 LA pulses) with the open-cell setup, we determined LODs for REEs in the ranges from 0.04 to 0.3 ng/g and from 0.08 to 0.5 ng/g for the glass NIST SRM 612 and the stalagmite sample, respectively. However, evaluation of time traces on a single point basis, as required to establish spatially resolved trace element profiles in the stalagmite, would result in approximately 35 times increased LODs. This is a result of the shorter integration period.

Jochum et al. [22] determined LODs for REEs in speleothems using a 213 nm LA system coupled to the Element 2 SF-ICPMS with the standard-interface configuration. LA was performed in a closed ablation chamber with a 110 µm diameter circular

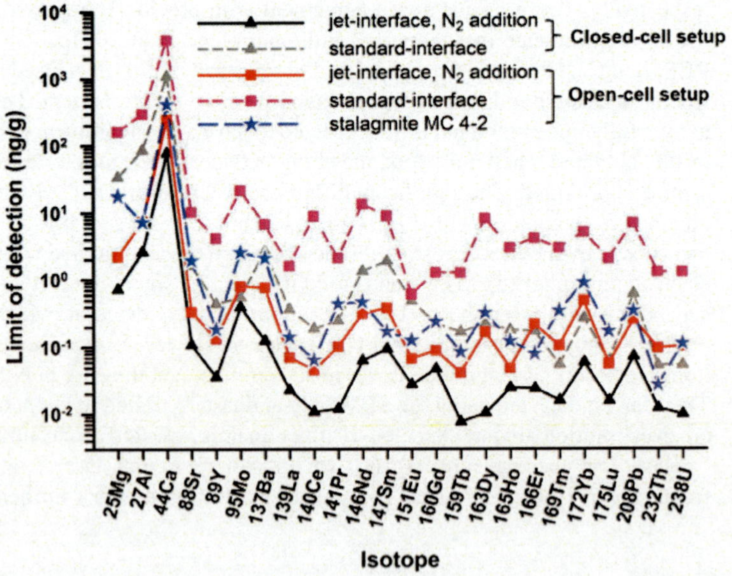

Fig. 4.4 Limits of detection (LODs) obtained for NIST SRM 612 and the stalagmite sample using a 160 µm diameter circular laser spot, 10 Hz ablation frequency and a scan speed of 10 µm/s. LODs are based on counting statistics [7]. Reprinted from Wu et al. [45] with permission of copyright Elsevier

laser spot and 10 Hz ablation frequency [22]. LODs were in the range from 0.5 to 5 ng/g. These calculations were based on the conventional 3σ approach proposed by Longerich et al. [27]. Following this approach would lower the LOD estimates for the experiments reported in this study by approximately one order of magnitude in cases of low background signal intensities.

SF-ICPMS typically provides high ion transmission and usually low instrumental background, which translates into low LODs. However, magnetic scanning limits the scan speed and therefore the isotope coverage when short transient signals need to be registered. The open-cell LA approach can also interface with other types of ICPMS instruments, such as Q-ICPMS or TOF-ICPMS, while the jet-interface is currently available for the sector field instruments of Thermo Fisher.

4.3.4 Evaluation of Accuracy and Precision

Figure 4.5 reports recoveries calculated for the analysis of USGS MACS-3. Here, the recovery was defined as the ratio between the quantified concentration for a given element and its reference concentration reported in the GeoReM database [21]. Recoveries are in the range from 0.9 to 1.1 for most of the elements studied. Apart from Ho, the quantified REE concentrations agreed with the USGS MACS-3

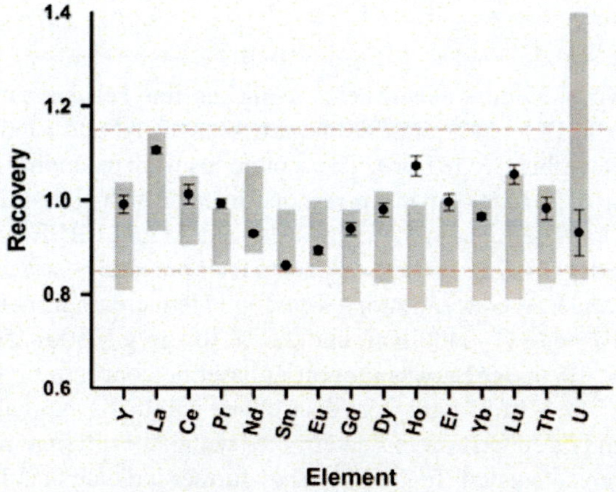

Fig. 4.5 Recoveries determined for the selected elements in USGS MACS-3. The experiment has been carried out with the open-cell LA setup coupled to the jet-interface equipped SF-ICPMS. LA was performed for at least 30 s with a 160 μm diameter circular laser spot, 10 Hz ablation frequency and 10 μm/s scan speed. The error bars represent the standard deviation calculated from four replicate measurements. The grey bars indicate the range of concentrations reported on the GeoReM database [21]. The dashed red lines show the ±15% interval of the GeoReM preferred data. Reprinted from Wu et al. [45] with permission of copyright Elsevier

reference values when the respective uncertainties were taken into account. Quantitative results of the analyses of USGS MACS-3, NIST SRM 612 and NIST SRM 614 are reported in Table S1. Relative deviations from reference concentrations were typically smaller than 12%.

In these experiments, the internal precision and external precision of the open-cell LA-SF-ICPMS approach were evaluated. Internal precision was defined as the variability of the element concentrations over a single measurement. External precision was defined as the variability of the element concentrations of repeat measurements.

From LA of USGS MACS-3, internal precision ranged from 0.2 to 1.7% (one standard error of the mean, 1 σ_m). External precision calculated from data of twenty replicate measurements carried out over the course of half a month was in the range from 2.6 to 10% (one relative standard deviation, 1 RSD) (Table 4.S2). As an example, the external uncertainty of the La concentration in USGS MACS-3 was 3.0%, 2–13 times larger than internal one (Fig. 4.S3). The higher external uncertainties could be a result of time-dependent instrumental mass discrimination that cannot be fully corrected for or potential heterogeneity of the pressed powder pellet. The high internal precision demonstrated that this open-cell LA-SF-ICPMS approach is suitable to determine high-resolution element concentrations profiles by evaluating time traces of signal intensities on a single point basis.

4.3.5 Microanalytical Assessment of a Stalagmite Sample

Figure 4.6 shows concentration profiles for selected elements. The profiles calculated from two individual experiments using the line scanning approach were in good agreement. Distinct anomalies in particular for Al and LREE concentrations were observed within two regions. These intervals were extending from 2700 to 2900 μm and from 4500 to 5000 μm, corresponding to growth periods of 79.7 to 82.8 and 82.8 to 86.3 ka. These anomaly regions showed Al and LREE concentrations that were enhanced by 2–17 times with respect to their pre- and post-anomaly region baselines. For example, Al concentrations were found to increase from baseline levels between 5.3×10^3 ng/g (1–2400 μm) and 2.5×10^3 ng/g (5300–7500 μm) to peak values of 1.3×10^5 ng/g. La was present at baseline concentrations of 17–18 ng/g while concentrations of 87 ng/g were determined within the anomaly regions. Box plots of the Al and La concentrations determined within the different regions of the stalagmite have been calculated (Fig. 4.S4). They further indicate that differences in Al and La concentrations between the anomaly and pre-/post-anomaly regions are significant.

In addition to the line scanning LA experiments that produce spatially resolved concentration profiles, the elemental composition of one anomaly region (≈4800 μm) and the adjacent baseline regions (≈5800 μm and ≈3800 μm) was also analyzed in single hole drilling experiments. For La, we determined a baseline concentration of 12–13 ng/g while a concentration of 295 ng/g was determined within the anomaly layer. The differences observed between anomaly region concentrations determined using the line scanning and the single hole drilling approach can

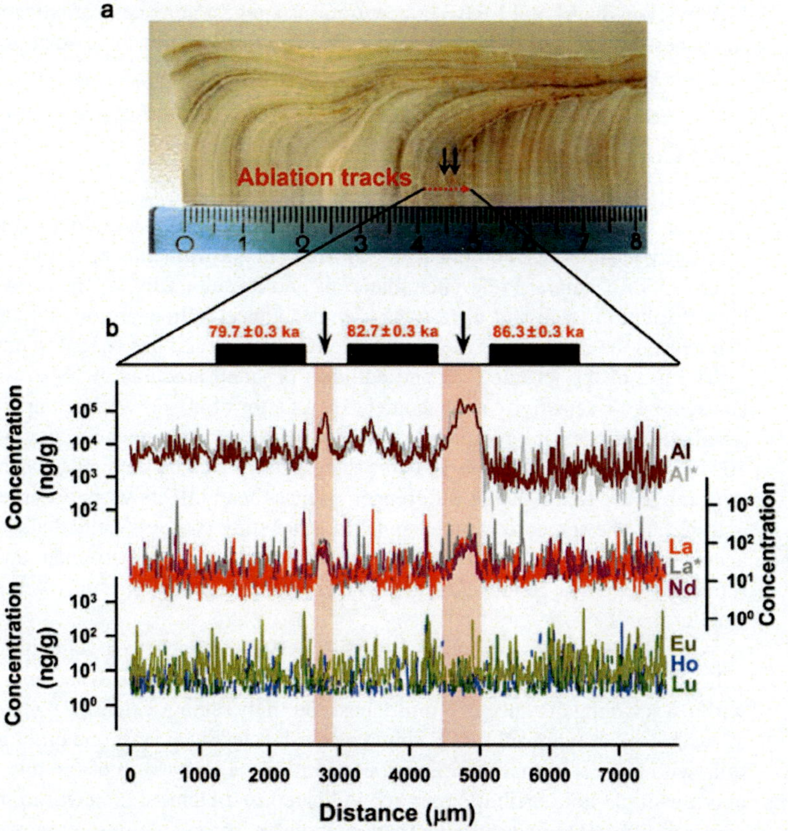

Fig. 4.6 **a** A photograph of an 8 cm segment of the stalagmite sample MC 4-2. The red dotted line indicates the position of the ablation tracks. The two black arrows point at the regions with elevated LREE contents. **b** Al and REE concentration profiles obtained by scanning along the red dotted line. Al* and La* profiles have been obtained from a second line scanning ablation paralleling the first one. The black horizontal bars denote locations where subsamples for U-Th dating were taken from. Al and LREE profiles show two pronounced anomalies at distances of 2700 μm to 2900 μm and 4500 μm to 5000 μm. These intervals correspond to the time periods of 79–82 and 82–86 ka (pink vertical bars). Reprinted from Wu et al. [45] with permission of copyright Elsevier

be attributed to the different sampling methods and the finite size of the anomaly region. While the single hole drilling approach ideally samples a constant fraction of the anomaly region, the line scanning continuously traverses this region and the aerosol dispersion essentially dilutes material with higher trace element contents. In general, the concentrations of major, minor and trace elements determined in hole drilling experiments are comparable with the ones obtained from line scanning studies.

Elevations in Al and LREE concentrations suggest an excess of siliceous rocks and ashes. Volcanic eruptions might be the main contributors to these anomalies.

4.4 Conclusions

An open-cell LA setup with a gas exchange device (GED) was coupled to a high sensitivity jet-interface equipped SF-ICPMS. This instrumental approach was characterized with standard reference materials and its suitability for the acquisition of high-resolution trace and ultra-trace element concentration profiles in stalagmites that exceed the dimensions of a conventional ablation cell has been demonstrated.

The use of the jet-interface and addition of small amounts of N_2 to the carrier gas achieved a sensitivity enhancement by a factor of 20–40 when compared to the standard interface configuration. From LA with 160 μm diameter circular laser spots, 10 Hz ablation frequency and 10 μm/s scan speed, we could obtain LODs in the sub-ng/g range for REEs. Little differences in mass load effects were observed for Ca and REEs. The standard sample standard bracketing strategy yielded a satisfactory accuracy for trace and ultra-trace element determinations in carbonate and silicate reference materials. Relative deviations from reference concentrations were typically smaller than 12%.

Application of this open-cell LA-SF-ICPMS approach to a stalagmite sample collected from East Timor revealed two abrupt elevations in LREE concentrations within a scanning distance of 8 mm. These anomaly regions extended over a distance of ≈200 μm and showed LREE abundances elevated by at least one order of magnitude with respect to their baseline concentrations. Comparison of the line scanning and the single hole drilling approach indicated that element concentration profiles can readily be obtained in speleothem carbonates. A notable limitation of the scanning approach with the open-cell LA setup was the occurrence of signal spikes, which most likely arose from remobilized particles. Nevertheless, this work highlights the potential and the applicability of our high-sensitivity open-cell LA-ICPMS methodology for paleo-climatic and paleo-environmental studies.

References

1. Akagi T, Hashimoto Y, Fu FF, Tsuno H, Tao H, Nakano Y (2004) Variation of the distribution coefficients of rare earth elements in modern coral-lattices: species and site dependencies. Geochim Cosmochim Acta 68(10):2265–2273
2. Baker J, Waight T, Ulfbeck D (2002) Rapid and highly reproducible analysis of rare earth elements by multiple collector inductively coupled plasma mass spectrometry. Geochim Cosmochim Acta 66(20):3635–3646
3. Borovinskaya O, Gschwind S, Hattendorf B, Tanner M, Günther D (2014) Simultaneous mass quantification of nanoparticles of different composition in a mixture by microdroplet generator-ICPTOFMS. Anal Chem 86(16):8142–8148

4. Bourdin C, Douville E, Genty D (2011) Alkaline-earth metal and rare-earth element incorporation control by ionic radius and growth rate on a stalagmite from the Chauvet Cave Southeastern France. Chem Geol 290(1–2):1–11
5. Burger M, Gundlach-Graham A, Allner S, Schwarz G, Wang HA, Gyr L et al (2015) High-speed, high-resolution, multielemental LA-ICP-TOFMS imaging: Part II. Critical evaluation of quantitative three-dimensional imaging of major, minor, and trace elements in geological samples. Anal Chem 87(16):8259–8267
6. Cruz FW Jr, Burns SJ, Jercinovic M, Karmann I, Sharp WD, Vuille M (2007) Evidence of rainfall variations in Southern Brazil from trace element ratios (Mg/Ca and Sr/Ca) in a Late Pleistocene stalagmite. Geochim Cosmochim Acta 71(9):2250–2263
7. Currie LA (1968) Limits for qualitative detection and quantitative determination. Application to radiochemistry. Anal Chem 40(3):586–593
8. Fairchild IJ, Baker A, Borsato A, Frisia S, Hinton RW, McDermott F, Tooth AF (2001) Annual to sub-annual resolution of multiple trace-element trends in speleothems. J Geol Soc 158(5):831–841
9. Fairchild IJ, Smith CL, Baker A, Fuller L, Spötl C, Mattey D, McDermott F (2006) Modification and preservation of environmental signals in speleothems. Earth Sci Rev 75(1–4):105–153
10. Fairchild IJ, Treble PC (2009) Trace elements in speleothems as recorders of environmental change. Quatern Sci Rev 28(5–6):449–468
11. Fallon SJ, White JC, McCulloch MT (2002) Porites corals as recorders of mining and environmental impacts: Misima Island, Papua New Guinea. Geochim Cosmochim Acta 66(1):45–62
12. Fleitmann D, Burns SJ, Mudelsee M, Neff U, Kramers J, Mangini A, Matter A (2003) Holocene forcing of the Indian monsoon recorded in a stalagmite from southern Oman. Science 300(5626):1737–1739
13. Fricker MB (2012) Design of ablation cells for LA-ICP-MS: from modeling to high spatial resolution analysis applications (Doctoral dissertation, ETH Zurich)
14. Fryer BJ, Jackson SE, Longerich HP (1995) The design, operation and role of the laser-ablation microprobe coupled with an inductively coupled plasma; mass spectrometer (LAM-ICP-MS) in the earth sciences. Can Mineral 33(2):303–312
15. Grottoli AG, Matthews KA, Palardy JE, McDonough WF (2013) High resolution coral Cd measurements using LA-ICP-MS and ID-ICP-MS: calibration and interpretation. Chem Geol 356:151–159
16. Günther D, Hattendorf B (2005) Solid sample analysis using laser ablation inductively coupled plasma mass spectrometry. TrAC Trends Anal Chem 24(3):255–265
17. Hattendorf B (2002) Ion molecule reactions for the suppression of spectral interferences in elemental analysis by inductively coupled plasma mass spectrometry (Doctoral dissertation, ETH Zurich)
18. Hattendorf B, Günther D (2014) Laser Ablation Inductively Coupled Plasma Mass Spectrometry (LA-ICPMS). In: Handbook of spectroscopy. Second, Enlarged Edition, pp 647–698
19. Hori M, Ishikawa T, Nagaishi K, You CF, Huang KF, Shen CC, Kano A (2014) Rare earth elements in a stalagmite from southwestern Japan: a potential proxy for chemical weathering. Geochem J 48(1):73–84
20. Hu Z, Liu Y, Gao S, Liu W, Zhang W, Tong X et al (2012) Improved in situ Hf isotope ratio analysis of zircon using newly designed X skimmer cone and jet sample cone in combination with the addition of nitrogen by laser ablation multiple collector ICP-MS. J Anal At Spectrom 27(9):1391–1399
21. Jochum KP, Nohl U, Herwig K, Lammel E, Stoll B, Hofmann AW (2005) GeoReM: a new geochemical database for reference materials and isotopic standards. Geostand Geoanal Res 29(3):333–338
22. Jochum KP, Scholz D, Stoll B, Weis U, Wilson SA, Yang Q et al (2012) Accurate trace element analysis of speleothems and biogenic calcium carbonates by LA-ICP-MS. Chem Geol 318:31–44

23. Kawabe I, Kitahara Y, Naito K (1991) Non-chondritic yttrium/holmium ratio and lanthanide tetrad effect observed in pre-cenozoic limestones. Geochem J 25(1):31–44
24. Kivel N, Potthast HD, Günther-Leopold I, Vanhaecke F, Günther D (2014) Modeling of the plasma extraction efficiency of an inductively coupled plasma-mass spectrometer interface using the direct simulation Monte Carlo method. Spectroc. Acta Pt B-Atom Spectr 93:34–40
25. Koch J, Günther D (2011) Review of the state-of-the-art of laser ablation inductively coupled plasma mass spectrometry. Appl Spectrosc 65(5):155–162
26. Kroslakova I, Günther D (2007) Elemental fractionation in laser ablation-inductively coupled plasma-mass spectrometry: evidence for mass load induced matrix effects in the ICP during ablation of a silicate glass. J Anal At Spectrom 22(1):51–62
27. Longerich HP, Jackson SE, Günther D (1996) Inter-laboratory note. Laser ablation inductively coupled plasma mass spectrometric transient signal data acquisition and analyte concentration calculation. J Anal At Spectrom 11(9): 899–904
28. McDermott F (2004) Palaeo-climate reconstruction from stable isotope variations in speleothems: a review. Quatern Sci Rev 23(7–8):901–918
29. Mertz-Kraus R, Brachert TC, Jochum KP, Reuter M, Stoll B (2009) LA-ICP-MS analyses on coral growth increments reveal heavy winter rain in the Eastern Mediterranean at 9 Ma. Palaeogeogr Palaeoclimatol Palaeoecol 273(1–2):25–40
30. Richter DK, Gotte T, Niggemann S, Wurth G (2004) REE3+ and Mn2+ activated cathodoluminescence in lateglacial and Holocene stalagmites of central Europe: evidence for climatic processes? Holocene 14(5):759–768
31. Scherer M, Seitz H (1980) Rare-earth element distribution in Holocene and Pleistocene corals and their redistribution during diagenesis. Chem Geol 28:279–289
32. Shen CC, Wu CC, Liu Y, Yu J, Chang CC, Lam DD et al (2011) Measurements of natural carbonate rare earth elements in femtogram quantities by inductive coupled plasma sector field mass spectrometry. Anal Chem 83(17):6842–6848
33. Shen CC, Wu CC, Cheng H, Edwards RL, Hsieh YT, Gallet S et al (2012) High-precision and high-resolution carbonate ^{230}Th dating by MC-ICP-MS with SEM protocols. Geochim Cosmochim Acta 99:71–86
34. Sholkovitz E, Shen GT (1995) The incorporation of rare earth elements in modern coral. Geochim Cosmochim Acta 59(13):2749–2756
35. Siklósy Z, Demeny A, Vennemann TW, Pilet S, Kramers J, Leél-Őssy S et al (2009) Bronze Age volcanic event recorded in stalagmites by combined isotope and trace element studies. Rapid Commun Mass Sp 23(6):801–808
36. Tabersky D, Nishiguchi K, Utani K, Ohata M, Dietiker R, Fricker MB et al (2013) Aerosol entrainment and a large-capacity gas exchange device (Q-GED) for laser ablation inductively coupled plasma mass spectrometry in atmospheric pressure air. J Anal At Spectrom 28(6):831–842
37. Tan L, Shen CC, Cai Y, Lo L, Cheng H, An Z (2014) Trace-element variations in an annually layered stalagmite as recorders of climatic changes and anthropogenic pollution in Central China. Quatern Res 81(2):181–188
38. Tanaka K, Takahashi Y, Shimizu H (2007) Determination of rare earth element in carbonate using laser-ablation inductively-coupled plasma mass spectrometry: an examination of the influence of the matrix on laser-ablation inductively-coupled plasma mass spectrometry analysis. Anal Chim Acta 583(2):303–309
39. Taylor SR, McLennan SM (1981) The composition and evolution of the continental crust: rare earth element evidence from sedimentary rocks. Phil Trans R Soc Lond A 301(1461):381–399
40. Treble P, Shelley JMG, Chappell J (2003) Comparison of high resolution sub-annual records of trace elements in a modern (1911–1992) speleothem with instrumental climate data from southwest Australia. Earth Planet Sci Lett 216(1–2):141–153
41. Treble PC, Chappell J, Shelley JMG (2005) Complex speleothem growth processes revealed by trace element mapping and scanning electron microscopy of annual layers. Geochim Cosmochim Acta 69(20):4855–4863

42. Wang YJ, Cheng H, Edwards RL, An ZS, Wu JY, Shen CC, Dorale JA (2001) A high-resolution absolute-dated late Pleistocene monsoon record from Hulu Cave, China. Science 294(5550):2345–2348
43. Webb GE, Kamber BS (2000) Rare earth elements in Holocene reefal microbialites: a new shallow seawater proxy. Geochim Cosmochim Acta 64(9):1557–1565
44. White WB (2007) Cave sediments and paleoclimate. J Cave Karst Stud 69(1):76–93
45. Wu C-C, Burger M, Günther D, Shen C-C, Hattendorf B (2018) Highly-sensitive open-cell LA-ICPMS approaches for the quantification of rare earth elements in natural carbonates at parts-per-billion levels. Anal Chim Acta 1018:54–61
46. Wyndham T, McCulloch M, Fallon S, Alibert C (2004) High-resolution coral records of rare earth elements in coastal seawater: biogeochemical cycling and a new environmental proxy3. Geochim Cosmochim Acta 68(9):2067–2080
47. Zhou H, Wang Q, Zhao J, Zheng L, Guan H, Feng Y, Greig A (2008) Rare earth elements and yttrium in a stalagmite from Central China and potential paleoclimatic implications. Palaeogeogr Palaeoclimatol Palaeoecol 270(1–2):128–138

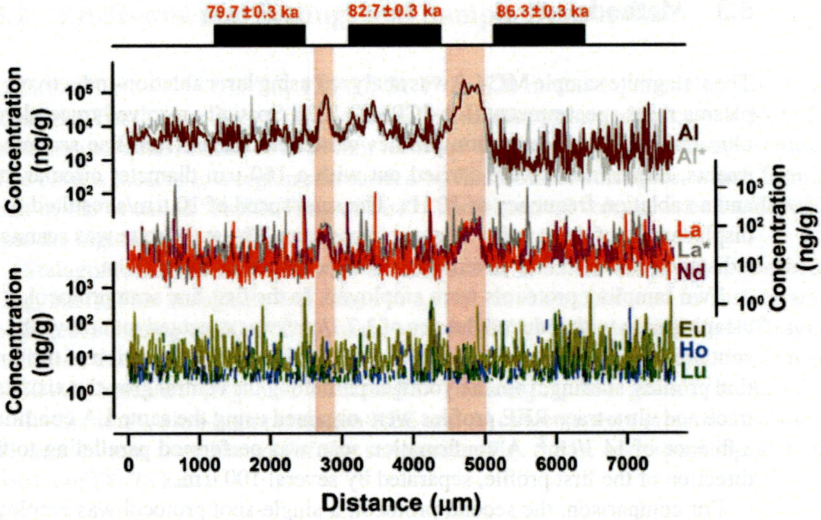

Fig. 5.2 Al and REE profiles obtained by a line scan (color curves) and a parallel confirmation scan (grey, labeled with star) using open-cell LA-ICPMS. Three horizontal bars denote depth intervals for U-Th dating. Al and REE profiles show two main and pronounced events at distance of 2700–2900 and 4500–5000 μm, corresponding to 79–82 and 82–86 ka (pink vertical bars). Adapted and reproduced from Wu et al. [23] with permission of copyright Elsevier

The post-Archean Australian shale (PAAS) [13] normalized REE diagram of the anomaly region (≈4800 μm) has been calculated from data recorded following the single hole drilling protocol. It is characterized by a flat pattern with a positive Eu anomaly. This pattern shows a similar trend as the volcano Toba caldera [22] and andesite leachate [8] (Fig. 5.3). Elevations of REE concentrations suggest an excess of siliceous rocks and ashes. Volcanic eruptions might be the main contributors to these anomalies.

5.4.2 Aluminum

Aluminum is one of the characteristic and abundant elements in igneous rocks, generally contains mass fraction of 10^3–10^4 μg/g [14]. Al has been used as a trace indicator for volcanic events recorded in carbonates [21, 24]. Two pronounced Al pulses, found with values of $1.3 * 10^5$ ng/g and $8.7 * 10^4$ ng/g, 14–22 times higher than the baseline values of $6.5 * 10^2$–$5.9 * 10^3$ ng/g. The concurrent increase of REE and All contents in the same time interval (Fig. 5.2), can be explained by the same igneous source.

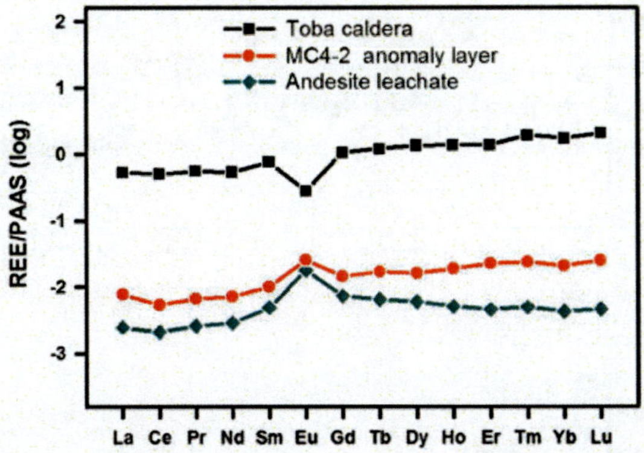

Fig. 5.3 Comparison of shale-normalized [13]. REE patterns of stalagmite MC 4-2, andesite leachate and Toba volcano caldera. The normalized REE pattern of the stalagmite MC 4-2 has been calculated from data acquired in a hole drilling experiment. This measurement was carried out within the anomaly region (≈4800 μm in Fig. 5.2). Reprinted from Wu et al. [23] with permission of copyright Elsevier

5.5 Oxygen and Carbon Isotopes

REEs and Al show evidence of the possible volcanic eruption events recorded in a stalagmite of Lekiraka Cave; however, it is unclear whether the eruptions had any climatic and environmental impacts in East Timor. $\delta^{13}C$ and $\delta^{18}O$ values in stalagmite are well-established climatic and ecological proxies [5]. $\delta^{13}C$ responds sensitively to climate induced changes in the type and density of vegetation and soil microbial activity. $\delta^{18}O$ variations are most likely closely to changes in temperature and seasonality of precipitation [6].

Here $\delta^{13}C$ and $\delta^{18}O$ values show no significant fluctuations following two REE anomalies (Fig. 5.4c,d), indicating that the eruptions had no detectable effect on local climate and vegetation. Although ≈0.5 mm (≈4500–5000 μm) and ≈0.2 mm (≈2700–2900 μm) of tephra was found in stalagmite MC 4-2, vegetation above Lekiraka Cave was not directly affected by the ash fall, based on its minor $\delta^{13}C$ variations.

5.6 Conclusions

REEs and aluminum measured on a well-dated section of stalagmite MC 4-2 from Lekiraka Cave in East Timor real two possible tephra fingerprints in East Timor between 88 and 78 ka. Two prominent elevations in LREE are in good agreement with Al anomalies, indicate that a combination of REE and Al appear to be a suitable proxy for tracing volcanic eruptions. Stable isotope $\delta^{13}C$ and $\delta^{18}O$ profiles display no

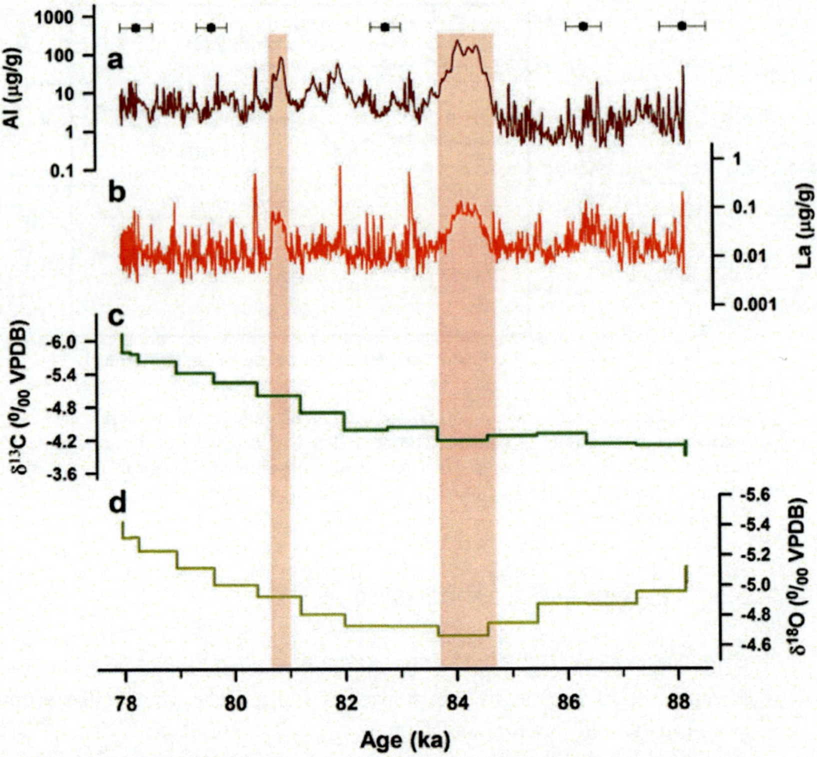

Fig. 5.4 Trace element and stable isotope profiles of the stalagmite MC 4-2. **a** Aluminum (dark red) profiles. **b** La (red) profiles. **c** $\delta^{13}C$ (green) and **d** $\delta^{18}O$ data (yellow). Vertical pink bars denote two abrupt events of REE and Al between 79 to 82 and 82 to 86 ka

profound response following the REE anomalies, suggesting that such two possible eruptions had no significant or detectable effect on climate and environment in East Timor. Since stalagmites can offer a superior dating accuracy [7] that utilizing such tephra proxy has greatest potential for improving existing tephrachronologies derived from less well-dated sediment records.

References

1. Badertscher S, Borsato A, Frisia S, Cheng H, Edwards RL, Tüysüz O, Fleitmann D (2014) Speleothems as sensitive recorders of volcanic eruptions–the Bronze Age Minoan eruption recorded in a stalagmite from Turkey. Earth Planet Sci Lett 392:58–66
2. Campbell IH, Czamanske GK, Fedorenko VA, Hill RI, Stepanov V (1992) Synchronism of the siberian traps and the permian-triassic boundary. Science 258(5089):1760–1763

3. Cole-Dai J (2010) Volcanoes and climate. Wiley Interdisciplinary Rev Clim Change 1(6):824–839
4. DeLong KL, Quinn TM, Taylor FW, Lin K, Shen CC (2012) Sea surface temperature variability in the southwest tropical Pacific since AD 1649. Nat Clim Chang 2(11):799
5. Fairchild IJ, Smith CL, Baker A, Fuller L, Spötl C, Mattey D, McDermott F (2006) Modification and preservation of environmental signals in speleothems. Earth Sci Rev 75(1–4):105–153
6. Fleitmann D, Cheng H, Badertscher S, Edwards RL, Mudelsee M, Göktürk OM et al (2009) Timing and climatic impact of Greenland interstadials recorded in stalagmites from northern Turkey. Geophysi Res Lett 36(19)
7. Henderson GM (2006) Caving in to new chronologies. Science 313(5787):620–622
8. Hori M, Ishikawa T, Nagaishi K, You CF, Huang KF, Shen CC, Kano A (2014) Rare earth elements in a stalagmite from southwestern Japan: a potential proxy for chemical weathering. Geochem J 48(1):73–84
9. Huang CY, Zhao M, Wang CC, Wei G (2001) Cooling of the South China Sea by the Toba eruption and correlation with other climate proxies~ 71,000 years ago. Geophys Res Lett 28(20):3915–3918
10. Kennett JP, McBirney AR, Thunell RC (1977) Episodes of Cenozoic volcanism in the circum-Pacific region. J Volcanol Geoth Res 2(2):145–163
11. Kutterolf S, Jegen M, Mitrovica JX, Kwasnitschka T, Freundt A, Huybers PJ (2013) A detection of Milankovitch frequencies in global volcanic activity. Geology 41(2):227–230
12. Mason BG, Pyle DM, Dade WB, Jupp T (2004) Seasonality of volcanic eruptions. J Geophysi Res 109(B4)
13. McLennan SM (1989) Rare earth elements in sedimentary rocks; influence of provenance and sedimentary processes. Rev Mineral Geochem 21(1):169–200
14. Nockolds SR (1954) Average chemical compositions of some igneous rocks. Geol Soc Am Bull 65(10):1007–1032
15. Rampino MR, Self S (1992) Volcanic winter and accelerated glaciation following the Toba super-eruption. Nature 359(6390):50
16. Renne PR, Basu AR (1991) Rapid eruption of the Siberian Traps flood basalts at the Permo-Triassic boundary. Science 253(5016):176–179
17. Robock A (2000) Volcanic eruptions and climate. Rev Geophys 38(2):191–219
18. Shen CC, Lin K, Duan W, Jiang X, Partin JW, Edwards RL et al (2013) Testing the annual nature of speleothem banding. Sci Rep 3:2633
19. Shen CC, Wu CC, Cheng H, Edwards RL, Hsieh YT, Gallet S et al (2012) High-precision and high-resolution carbonate ^{230}Th dating by MC-ICP-MS with SEM protocols. Geochim Cosmochim Acta 99:71–86
20. Stothers RB (1984) The great Tambora eruption in 1815 and its aftermath. Science 224(4654):1191–1198
21. Siklósy Z, Demeny A, Vennemann TW, Pilet S, Kramers J, Leél-Őssy S et al (2009) Bronze Age volcanic event recorded in stalagmites by combined isotope and trace element studies. Rapid Commun Mass Spectrom 23(6):801–808
22. Westgate JA, Shane PA, Pearce NJ, Perkins WT, Korisettar R, Chesner CA et al (1998) All Toba tephra occurrences across peninsular India belong to the 75,000 yr BP eruption. Quatern Res 50(1):107–112
23. Wu C-C, Burger M, Günther D, Shen C-C, Hattendorf B (2018) Highly-sensitive open-cell LA-ICPMS approaches for the quantification of rare earth elements in natural carbonates at parts-per-billion levels. Anal Chim Acta 1018:54–61
24. Wu C-C, Shen C-C, Lo L, Hsin Y-C, Yu K, Chang C-C et al (2018) Pinatubo volcanic eruption exacerbated an abrupt coral mortality event in 1991 summer. Geophys Res Lett 45:12396–12402

Chapter 6
Summary and Perspectives

In this dissertation, we have addressed two state-of-the-art techniques to analyze ultra-trace rare earth elements (REEs) in natural carbonates using SN-ICPMS and LA-ICPMS. Here the summary and the perspectives are described as below:

The first protocol demonstrates a direct analysis without chemical separation steps for rare earth elements (REEs) in carbonate samples with femtogram quantity has been successfully established at the HISPEC laboratory, NTU (*Analytical Chemistry*, *83*, 6842–6848, 2011). This approach promises a high-precision analysis that internal precisions of 0.2–1.0% (2σ) and external precisions of 1.9–6.5% (2σ) can be achieved. This high sample throughput, high precision and high temporal resolution REE technique has the potential to natural slow growth-rate carbonate materials, such as sclerosponges, tufa, foraminifera, and speleothems, and could provide valuable time-series records to advance our understanding of paleoclimatic and paleoenvironmental dynamics on different timescales.

Applied this technique to coral samples collected from the South China Sea region, monthly REE records of two modern corals register the 1991 Mount Pinatubo volcanic eruption (*Geophysical Research Letters*, *45,* 12396–12402. 2018). The results reveal that abrupt and extensive mortality episodes, such as which occurred in the South China Sea (SCS) in mid-1991, cannot be explained solely on the basis of high summer sea surface temperatures. The second largest volcanic eruption of the twentieth century, the 1991 Mount Pinatubo volcanic eruption, exacerbated this coral mortality event. The sensitive impacts by such a volcanic eruption on modern vulnerable coral reefs, associated with the stress of thermal conditions are demonstrated.

The second protocol, a high-sensitivity open-cell LA-ICPMS technique allows analyses of carbonate REE concentration profiles above a level of digit parts-per-billion (*Analytica Chimica Acta*, *1018,* 54–61, 2018). With an improved setup by adopting the "jet" vacuum interface, the sensitivity could be enhanced by 20–40 times compared to the conventional interface. Limits of detection reach as low as sub-ng/g range for most ultra-trace elements and the accuracy is better than 12% at

C.-C. Wu, *Advanced and Applied Studies on Ultra-Trace Rare Earth Elements (REEs) in Carbonates Using SN-ICPMS and LA-ICPMS*, Springer Theses,
https://doi.org/10.1007/978-981-16-3619-6_6

Table A Coral ST0506 REE/Ca (nmol/mol) and La/Yb contents

Corresponding Date (CE)	La/Ca	Ce/Ca	Pr/Ca	Nd/Ca	Sm/Ca	Eu/Ca	Gd/Ca	Tb/Ca	Dy/Ca	Ho/Ca	Er/Ca	Tm/Ca	Yb/Ca	Lu/Ca	La/Yb
1990.150	21.22	19.29	3.771	15.94	3.705	0.8394	6.648	0.9883	6.335	1.311	4.367	0.6664	4.802	0.7970	4.419
1990.225	22.86	34.50	5.071	24.00	4.504	1.240	9.029	0.6896	7.787	1.434	4.971	0.8195	5.523	0.9458	4.139
1990.300	17.31	14.81	3.655	18.09	2.880	0.9684	8.389	0.5197	7.722	1.517	5.389	0.7761	5.188	0.8420	3.337
1990.375	22.08	16.72	3.799	16.55	3.525	0.8928	5.036	1.057	5.863	1.486	4.718	0.6999	4.873	0.7855	4.531
1990.479	22.19	20.82	3.875	16.73	3.182	0.9168	5.698	1.077	6.041	1.521	5.026	0.8034	5.539	0.8811	4.006
1990.583	24.26	20.18	4.085	18.15	3.604	0.8988	6.173	1.086	6.526	1.521	4.894	0.7632	4.999	0.7778	4.853
1990.637	29.03	24.99	4.851	20.16	4.300	1.233	6.447	1.366	7.594	1.788	5.622	0.8131	5.601	0.8878	5.183
1990.690	30.43	33.62	5.621	23.08	5.086	1.280	7.301	1.458	7.497	1.741	5.106	0.7260	5.046	0.8104	6.031
1990.797	32.14	42.92	6.007	25.38	5.034	1.455	6.925	1.338	7.780	1.813	5.378	0.7299	5.066	0.8047	6.344
1990.850	31.37	25.93	5.509	22.33	4.780	1.244	6.825	1.370	7.552	1.891	5.735	0.7897	5.322	0.8666	5.894
1990.925	39.78	63.49	8.659	36.57	8.270	1.991	12.97	1.458	11.12	2.208	7.103	0.9972	6.595	1.081	6.032
1991.020	16.09	24.20	3.546	19.66	3.798	1.131	7.980	0.407	7.279	1.284	4.611	0.7143	4.985	0.7827	3.228
1991.155	34.04	52.52	7.446	34.47	7.593	1.876	12.86	1.187	11.10	1.960	6.374	0.9056	5.964	0.8926	5.708
1991.321	77.82	169.0	17.84	74.01	15.90	3.591	19.99	3.068	16.44	3.020	8.945	1.262	7.464	1.091	10.43
1991.467	259.3	736.0	64.54	254.1	52.43	11.40	45.79	9.402	32.27	5.628	14.80	1.933	11.87	1.667	21.84
1991.592	438.9	1244	106.5	413.1	84.59	17.57	67.25	14.03	43.85	7.280	18.54	2.173	13.20	1.819	33.25
1991.697	678.1	1733	163.4	655.5	131.4	27.73	112.3	22.11	71.68	11.67	29.18	3.587	20.05	2.775	33.82
HIATUS															
1992.000	41.45	28.45	7.526	32.63	6.633	1.680	8.891	1.732	9.415	2.216	7.044	1.095	7.709	1.329	5.377
1992.167	33.37	21.13	5.880	24.49	5.543	1.391	8.362	1.518	8.937	2.061	6.622	1.015	7.210	1.238	4.628
1992.333	38.41	24.89	7.017	30.04	6.200	1.582	8.542	1.548	8.451	1.904	5.939	0.8400	5.553	0.9067	6.917

(continued)

(continued)

Corresponding Date (CE)	La/Ca	Ce/Ca	Pr/Ca	Nd/Ca	Sm/Ca	Eu/Ca	Gd/Ca	Tb/Ca	Dy/Ca	Ho/Ca	Er/Ca	Tm/Ca	Yb/Ca	Lu/Ca	La/Yb
1992.500	41.85	26.79	7.724	33.32	7.206	1.742	10.22	1.690	9.232	1.990	6.029	0.8410	5.467	0.8610	7.655
1992.625	33.63	22.39	6.283	25.59	5.537	1.341	8.329	1.317	7.216	1.569	4.623	0.6385	4.028	0.6549	8.349
1992.800	31.72	20.78	5.716	24.08	4.912	1.256	8.066	1.251	7.039	1.555	4.826	0.6713	4.601	0.7347	6.894
1992.900	27.60	19.50	5.074	20.73	4.501	1.148	7.246	1.219	6.852	1.550	4.797	0.7244	5.105	0.8526	5.406
1993.000	19.30	11.83	3.184	13.87	3.017	0.7565	5.766	0.9142	5.537	1.255	4.024	0.6124	4.496	0.7567	4.293
1993.050	22.55	15.49	4.044	17.51	3.631	0.9680	6.702	1.016	6.181	1.395	4.288	0.6410	4.613	0.7403	4.888
1993.100	19.45	15.26	3.537	14.29	3.130	0.8331	4.235	0.9160	4.751	1.186	3.802	0.5952	3.981	0.6605	4.886
1993.150	20.38	15.25	3.720	15.50	3.342	1.173	5.005	0.9742	5.555	1.346	4.369	0.6931	4.839	0.8218	4.212
1993.200	20.40	16.42	3.690	15.90	3.357	0.9903	4.769	1.007	5.683	1.298	4.221	0.6734	4.910	0.8443	4.155
1993.250	18.53	13.83	3.325	13.75	3.155	0.8534	4.235	0.8894	5.103	1.272	4.173	0.6683	4.658	0.7599	3.978
1993.300	21.22	18.58	3.876	17.23	3.242	0.8700	5.416	1.003	5.361	1.297	4.191	0.6728	4.597	0.7216	4.616
1993.350	21.65	18.23	3.830	15.98	3.461	0.9193	5.557	1.006	5.695	1.406	4.383	0.6969	5.010	0.8136	4.321
1993.400	19.77	19.42	3.566	22.53	3.035	0.759	5.579	0.9645	5.056	1.126	3.597	0.5267	3.621	0.6071	5.460
1993.450	25.08	20.09	4.547	18.83	4.168	1.032	6.938	1.191	6.509	1.503	4.526	0.7033	4.978	0.8556	5.038
1993.500	6.060	26.72	5.320	22.63	4.869	1.193	7.854	1.298	7.318	1.633	5.105	0.7570	5.085	0.8736	1.192
1993.571	6.092	22.16	5.060	21.16	4.426	1.142	7.154	1.222	6.480	1.508	4.569	0.6567	4.695	0.7488	1.298

Table B Coral ST0506 Ba/Ca (μmol/mol), Al/Ca (μmol/mol), and Sr/Ca (mmol/mol) contents

Corresponding Date (CE)	Ba/Ca	Al/Ca	Sr/Ca
1990.150	2.706	10.27	9.054
1990.225	2.897	33.39	9.061
1990.300	4.169	14.81	8.849
1990.375	3.042	20.62	8.870
1990.479	3.039	20.62	8.866
1990.583	3.287	7.954	8.862
1990.637	3.565	20.33	8.858
1990.690	4.197	46.06	8.899
1990.797	4.373	33.33	8.887
1990.850	4.235	20.36	9.019
1990.925	4.113	76.26	9.008
1991.020	3.464	25.00	9.126
1991.155	3.386	40.83	9.049
1991.321	3.431	103.4	8.992
1991.467	3.639	183.1	9.027
1991.592	3.749	378.4	8.979
1991.697	4.146	240.4	9.058
HIATUS			
1992.000	5.769	11.56	9.165
1992.167	5.007	23.77	9.152
1992.333	3.587	122.2	8.956
1992.500	3.546	37.34	8.846
1992.625	3.551	30.87	8.814
1992.800	4.636	22.44	9.016
1992.900	5.908	123.5	9.191
1993.000	6.043	6.672	9.198
1993.050	5.712	30.50	9.153
1993.100	5.495	48.91	9.082
1993.150	5.083	16.46	9.011
1993.200	5.101	73.30	8.973
1993.250	5.388	1.937	8.894
1993.300	5.314	12.22	8.847
1993.350	5.615	6.647	8.860
1993.400	5.274	7.155	8.842
1993.450	5.204	4.038	8.810
1993.500	5.603	6.273	8.783
1993.571	6.653	4.168	8.778

Table C Coral MJL01a REE/Ca (nmol/mol) and La/Yb contents

Corresponding Date (CE)	La/Ca	Ce/Ca	Pr/Ca	Nd/Ca	Sm/Ca	Eu/Ca	Gd/Ca	Tb/Ca	Dy/Ca	Ho/Ca	Er/Ca	Tm/Ca	Yb/Ca	Lu/Ca	La/Yb
1990.001	4.464	5.634	0.9537	3.388	0.8108	0.1970	1.429	0.1782	1.328	0.2859	1.009	0.1583	1.142	0.1808	3.908
1990.101	5.213	5.340	1.474	8.322	1.037	0.1851	1.474	0.1678	1.107	0.2727	0.7658	0.1168	0.8307	0.1267	6.275
1990.192	12.14	13.61	2.605	8.202	2.259	0.4275	3.028	0.4030	2.783	0.6428	2.178	0.3524	2.519	0.4142	4.819
1990.283	7.717	12.17	2.068	9.071	1.166	0.3210	2.672	0.2974	2.363	0.5792	2.145	0.3297	2.805	0.4883	2.751
1990.374	12.21	18.86	7.653	8.85	1.921	0.4831	3.753	0.4562	3.765	0.7122	2.708	0.4322	3.465	0.5929	3.524
1990.465	12.58	17.86	3.738	8.35	1.752	0.4840	4.400	0.5407	3.702	0.8114	2.923	0.4973	3.666	0.5969	3.432
1990.556	6.902	11.43	1.835	8.614	0.9506	0.2330	1.883	0.2539	1.802	0.3954	1.301	0.2036	1.818	0.3050	3.797
1990.647	6.154	10.01	4.625	11.69	0.8660	0.2270	1.703	0.2445	1.942	0.3918	1.424	0.1996	1.521	0.2926	4.047
1990.738	5.664	8.314	1.633	10.35	0.6445	0.1482	1.806	0.2133	1.468	0.3304	1.182	0.1893	1.469	0.2655	3.855
1990.829	6.079	8.974	1.727	7.440	0.5944	0.2224	1.595	0.2292	1.534	0.3593	1.227	0.1921	1.494	0.2536	4.069
1990.92	10.97	15.38	3.518	19.80	1.972	0.4996	2.297	0.3379	2.571	0.6263	1.946	0.2836	2.346	0.4156	4.676
1991.011	11.21	15.12	2.406	9.779	1.746	0.4208	3.186	0.3479	2.633	0.5873	1.981	0.3374	2.303	0.3970	4.868
1991.102	15.39	20.11	2.951	10.11	2.086	0.5456	4.455	0.4045	3.650	0.7620	2.606	0.4233	3.229	0.5577	4.765
1991.245	18.37	21.66	3.637	12.91	2.515	0.6022	4.246	0.4706	3.556	0.7549	2.398	0.3736	2.795	0.5028	6.572
1991.468	49.85	51.20	9.217	30.46	5.698	1.226	7.486	0.8625	5.993	1.117	3.607	0.5371	3.935	0.6314	12.67
1991.601	27.06	27.41	5.266	17.18	2.658	0.6547	3.266	0.4238	2.589	0.6100	1.911	0.3000	2.170	0.3873	12.47
1991.774	6.910	8.919	1.155	4.080	0.7975	0.2302	1.585	0.1873	1.388	0.3127	1.110	0.1870	1.518	0.2549	4.551
1991.917	6.740	8.061	1.143	3.877	0.9496	0.1949	1.688	0.1974	1.561	0.3087	1.116	0.1723	1.348	0.2466	5.000
1992.06	15.04	16.28	2.621	8.806	2.014	0.5736	3.526	0.4147	3.268	0.7053	2.204	0.3469	2.767	0.4766	5.437
1992.203	11.66	11.58	2.115	9.756	1.449	0.2870	2.143	0.2681	1.991	0.4508	1.581	0.2262	1.788	0.3078	6.521
1992.332	17.49	16.50	5.150	9.779	2.084	0.5012	3.360	0.4067	3.054	0.6350	2.260	0.3105	2.642	0.4319	6.620

(continued)

(continued)

Corresponding Date (CE)	La/Ca	Ce/Ca	Pr/Ca	Nd/Ca	Sm/Ca	Eu/Ca	Gd/Ca	Tb/Ca	Dy/Ca	Ho/Ca	Er/Ca	Tm/Ca	Yb/Ca	Lu/Ca	La/Yb
1992.461	6.846	7.092	1.051	3.584	0.6913	0.1019	1.799	0.1332	1.368	0.2494	0.9707	0.1318	1.058	0.1608	6.472
1992.59	9.033	9.908	1.610	5.690	1.131	0.1963	2.495	0.2331	2.246	0.3884	1.460	0.2189	1.737	0.2901	5.202
1992.719	8.715	9.142	2.328	11.12	1.081	0.2935	1.937	0.2413	1.815	0.4093	1.395	0.2342	1.754	0.2873	4.970
1992.848	5.508	6.772	1.566	8.440	0.7680	0.1373	1.335	0.1749	1.389	0.3005	0.9877	0.1507	1.028	0.1869	5.360
1992.977	4.614	5.622	0.7415	2.669	0.4707	0.1050	1.416	0.1457	1.249	0.2342	0.7892	0.1456	0.9821	0.1666	4.698
1993.106	6.125	6.945	1.436	5.982	0.7430	0.2313	2.009	0.2161	1.666	0.3742	1.241	0.1745	1.363	0.2177	4.493
1993.231	6.598	8.178	1.333	4.196	1.040	0.2234	1.762	0.2327	1.575	0.3895	1.278	0.2084	1.650	0.2771	3.999
1993.356	5.919	7.476	1.465	6.006	0.9648	0.1983	1.923	0.1988	1.879	0.3600	1.292	0.2018	1.572	0.2727	3.764
1993.481	5.990	8.418	2.904	13.48	0.9910	0.1733	1.946	0.2132	2.087	0.3631	1.257	0.2090	1.673	0.2685	3.580